Robert Nelson

Asiatic Cholera

Its Origin and Spread

Robert Nelson

Asiatic Cholera
Its Origin and Spread

ISBN/EAN: 9783744762274

Printed in Europe, USA, Canada, Australia, Japan

Cover: Foto ©berggeist007 / pixelio.de

More available books at **www.hansebooks.com**

ASIATIC CHOLERA:

ITS ORIGIN AND SPREAD IN

ASIA, AFRICA, AND EUROPE,

INTRODUCTION INTO

AMERICA THROUGH CANADA;

REMOTE AND PROXIMATE CAUSES, SYMPTOMS AND PATHOLOGY,

AND THE

VARIOUS MODES OF TREATMENT ANALYZED.

BY R. NELSON, M.D.,

Health Commissioner during the first two invasions, 1832, 1834; President of
the Medical Board for the District of Montreal.

———

NEW YORK:

WILLIAM A. TOWNSEND, PUBLISHER,

434 BROOME STREET.

1866.

JOHN MEDOLE, Printer,
No. 193 *Pearl Street, N. Y.*

1

PREFACE.

It was through medical journals, commencing in 1820, that my attention was first drawn to the cholera of India, said to prevail in an epidemic form, and to its subsequent spread through the rest of Asia, the east of Africa and into Europe, even into the western confines of the latter continent. In the absence of personal observation, I became as thoroughly conversant with the malady as reading could afford information. But I could not help being struck with the great discrepancy in opinions and doctrines promulgated. On one point only all were in perfect accord; that point was one of patent fact—a description of the appearance of a patient and the phenomena manifested in every case. In all other respects they all disagreed, as doctors—doctors in medicine and doctors in theology—only can differ and argue.

Thus prepared by reading, and rendered skeptical by conflicting doctrines and opinions, I entered upon the duties of my office. Daily reports were sent in to the Board of Magistrates, and to my office as executive officer of the Medical Board, from practitioners, and other sources of information. It will be admitted that under these circumstances I was favorably placed to become well acquainted with the pest from its first appearance in America, in 1832, and the second invasion in 1834; nor have I remained unmindful of subsequent irruptions.

Of the contents of this book I have often spoken, for years past, with medical men, and as often have been urged to publish my views. I do so now, principally to fix a point in history:

The introduction of a new malady into a new country, and to record the experience I acquired in the beginning, and which has been confirmed by every case since then. Many of my statements come into direct collision with those of previous and even of present writers. The only apology I can offer for the difference is, mine are the result of personal observations carefully made on the patient, unbiassed by abstract theories, and an erroneous physiology, which possesses nothing in common with Asiatic cholera.

R. NELSON.

P. S. Since writing this book, some of my state-ments have met with corroboration this present sea-son, namely: 1st. Cholera broke out on board a French vessel a few days out of port. It soon had selected its victims, and then ceased; the remainder of the passage, which was of some weeks' duration, was completed in perfect health, when the vessel arrived at Guadaloupe, all well. 2d. An apparently healthy ship and passengers enter a new (land) atmosphere, and within a day or two the pest breaks out among the passengers who had escaped the pestilence on board, and communicate it to a healthy people, among whom it makes dreadful ravages. (See Sections 66, 67, 74, 87, 88.) 3d. Two steamers, from England, experience an outbreak on board a few days after leaving port. (Sec. 78.) But the vessels arrive with cholera still on board; the passage being of only two weeks' duration, the pest had not the usual period of time to select all the susceptible, and then cease. One put into Halifax, is placed in seclusion, and now, in a new (land) atmosphere, renews its activity, as has ever been the case. Some passengers escape from the quarantine inclosure, travel rapidly through a perfectly healthy country, take ill and die in the new atmosphere. These three vessels afford several other points corroborative of my statements, but which I do not notice, that I may not enlarge this book uselessly.

R. N.

INDEX.

ASIATIC CHOLERA.

CHAPTER I.

ORIGIN AND ITINERARY.

The 19th century is remarkable for the great events that have taken place since its commencement, and a short time previously. In it the knowledge of steam, its powers, the means of controlling it, its uses as a substitute for bodily and manual labor, and its complete obedience to the hand of man, have been perfected.

Chemistry, already in its infancy, has become an exact science.

Electricity and magnetism have advanced from being mere toys to a grade of the highest utility. Geology has disinterred the long buried almanac of the globe, brought clearly into light the reign of extinct creations, overwhelmed and hidden since millions of years have passed away. These are only a

2

few of the remarkable occurrences; but the one which distinguishes this century more than all others of which history makes mention, in relation to man, is the stupendous plague called Cholera. Stupendous from its wide spread malignancy over every continent; stupendous from the millions of victims it has swallowed; stupendous from the rapidity of its spread; stupendous from the few brief moments of life it allows to those it attacks: apparently capricious in its selections, it has desolated some places, spared others; terrified nations, arrested the march of armies, turned conquest into defeat, laughed at science, and baffled the efforts of man to arrest its empire.

1. Sporadic cases of cholera have been known in all historic ages, from that in which the great Hippocrates flourished until now, and is mentioned in all treatises of medicine. In India it has been long known. Bontius, Lind, Dellon, &c., almost all writers on the diseases of that country, make special mention of it. It is said to have prevailed in an epidemic form in the Bengalese territory in 1762, when it carried off 30,000 natives and 800 Europeans in one season, and then ceased.

2. A more recent account says that the Government of India became cognizant of an increase of the malady among the natives ten years, at least, before it assumed the character of terrible devastation that distinguished it in 1817; while, in all these ten years, not one case occurred among the European population, and during that period not one case was admitted into the General Hospital of Calcutta. It still maintained the simple and sporadic character noticed for many years; that is, single cases, here and there, now and then, happened apparently spontaneously and without assignable cause, not spreading from one to another individual, and still less existing as an epidemic.

3. The Government of Bengal was officially informed on the 28th of August, 1817, that a sudden, violent and spreading irruption of cholera had occurred in the populous town of Jessore, the capital of the Sunderbunds, situated in the delta of the Ganges, sixty miles from Calcutta; and that 10,000 of the inhabtitants had died in a few weeks. The report stated: " The inhabitants, astonished and terrified at the unaccountable and destructive pestilence, fled in crowds to the country as the only means of escaping impending death. So unforeseen and appalling was the attack, the public functionaries, in extreme consternation, closed the civil courts of the district, and business of every description was abandoned for a time. In the course of a few weeks 10,000 of the inhabitants perished in this single district."

It reached Calcutta on the 1st of September, whence it spread like radii from a centre to all points of the compass.

4. Although its origin is generally located at Jessore, it is asserted that it ravaged the banks of the Burrampooter early in June, reached Silket, in the east, and spread to the confines of Balasore and the provinces of Behar, in the east. To the northwest of Jessore it attacked Patna and Dinapore, July 11th. Following the course of the Ganges and its tributaries, it reached and attacked the camp of the Marquis of Hastings, in Bundelcund, on the banks of the Sinde, a tributary of the Jumna, on the 6th November. The invasion was so sudden and violent that mounted men were stricken from their steeds, fell and died on the road. The roads were encumbered with ill, dying and dead of the army, and its followers, more numerous about an Indian army than the army itself.   But this irruption, though astounding, was soon over; for, having commenced on the 6th November, no new cases occurred in the camp after the 8th December. However, the pestilence still followed in the track of the army, or rather the army sowed its seeds all along the roads, rivers, and ways of travel, infecting village and town in succession throughout Bundelcund, and spread to the provinces of Behar, Malwah, Candish, overshadowing nearly

all the Deccan. It killed 10,000 in the **town of**
Banda, and all the neighboring towns were similarly
oppressed. One of these towns, Kotah, lies on the
east side of the Chumbul, **and is** built upon a dry
rock; **but the** aridity of the location afforded no sanitary protection, since 100 persons perished daily;
the surviving inhabitants, struck at last with dismay,
fled the city.

5. Two offshoots of **the** pestilence appear to have
set out from the town of Saugur, the westerly one
going to Kotah, **and** the southern one, going due
south, attacked the forces stationed on the Nerbundah,
advanced through the states of Nagpore and Poonah,
passed through Seringapatan to the extreme point of
India. The **westerly** one from Kotah followed the
coast of Malabar. An eastern branch followed the
coast of Coromandel.

6. Each of these offsets invaded town after town,
village after village in succession, regardless of topography, high or low land, mountain, plain or swamp;
nor was the **high** Ghaut-range capable of presenting
a barrier to its destructive march.

7. Northwardly it reached Fyzabad and Lucknow
in April and May, 1818. At Gurruckpore it carried
off 30,000 of the inhabitants, while, during this appalling devastation, *not one* of the prisoners confined
in the jails took the pest. The high mountains that

separate Hindostan from Nepaul were invaded, as well as the high table land of Patna and Rhatgoun, 4,000 feet above the sea; Katamandoo, in Nepaul, was attacked in 1818, and also the south flank of those almost impassable heights; and in 1819 it broke out in Siam, Cochin China, beyond that range.

8. In October, 1818, Madras suffered, and thence cholera reached Candy, the capital of Ceylon, at a great height above the sea, in December. Commencing at the coast, it spread to the interior of the island.

9. We have now traced the course of the pestilence from its cradle in the Sunderbunds, eastwardly to the confines of Cochin China; northwardly to the Himmalayas; westwardly to Bombay; and southwardly to Cape Comorin, and into the island of Ceylon. Let us now follow its course beyond India over the sea into Sumatra, Borneo, the Moluccas, and the Philippines, into China.

10. Malacca, the extreme southern point of the Birman Empire, was attacked August, 1818; the pestilence was carried over to the island of Penang in the Straits; and in 1819 it spread through Sumatra. The town of Batavia alone lost 17,000, and the whole island 102,000, according to the Dutch statistics. Java suffered in April, 1821. The destruction in these places, like in Borneo, was appalling.

11. Manilla was attacked early in 1820, where the death of 15,000 victims alarmed the people to fury, who rose up against the foreigners and made a frightful massacre of them.

12. Traveling northwardly the cholera reached Canton in 1820. Bankok, in Siam, lost 40,000 in 1819. Nankin and Pekin suffered in 1823.

13. Southwesterly traveling across the sea it invaded the African islands, Mauritius and Island of Bourbon. The Topaz frigate, from Calcutta, anchored in Port St. Louis, Isle of France (Mauritius), at the end of November, 1819. Several of the crew had died of cholera on the passage. The captain refused to submit to quarantine; his people landed, infected the town, carrying off 20,000. (But the governor, Farquhar, perhaps to screen the captain of his haughty disobedience of the quarantine laws, reported the number at 7,000 or 8,000.) By strict measures adopted, the remainder of the island was spared.

14. Baron Melius, the French governor of the adjacent island, Bourbon, at once interdicted all communication with his colony. Still, in spite of severe but wise measures of the governor, the pestilence was brought there by the cupidity of a trading vessel, Pic-Var, which left Mauritius July 7, 1820. On the 14th cholera broke out at St. Denis. The inhabitants fled at once; the authorities surrounded the place with sentinels, kept up a most vigilant seclusion. Success was the result and the reward; the pest spread no further into the country; it was confined to the town, in which 259 were attacked, of whom 178 died.

15. August 6th, 1818, it reached Panwell, separated from Bombay by a strait crossed daily by boats. On the 9th a passenger from Panwell introduced the pestilence into Bombay and Salsette, whence it spread along the whole coast of Malabar.

16. By the extensive commerce of Bombay it was carried across the gulf to Muscat, in July, 1821. It soon broke out in the islands of Ormus and Kishme, at the entrance of the Persian Gulf; and by August it extended up the coast to Bender-Abassi, where the destruction was so great that the bazaars were closed, houses abandoned, and the dead lay in heaps unburied in the streets. September 6th it broke out at Shiraz, a hundred miles up the coast; here one of the first families attacked was that of the Prince Royal of Persia. One of his wives, his mother, and many Georgian beauties in his seraglio, besides some of his children, were carried off. The mortality was 6,000 out of a population of 35,000. He fled, with his harem and suite, from the fury of the pestilence.

17. Continuing its northwardly route it invaded

the towns of Yerd and Ispahan, where 25,000 fell its victims. Cachan, Koms and Teheran suffered, on its way to Tauris, in 1822. Boats carried it to Bassora and Bagdad, where it swept away one-third of the population. In July it infected Mosul, north of Bagdad, marched westward into Syria, and broke out at Aleppo, in November, 1822, spreading to Antioch, etc., along the coast of the Mediterranean to Tripoli.

18. Ascending the Caspian Sea it reached Astrachan, near the mouth of the Volga, in September, 1823, where the attacks and deaths were rapid. The Russian Government sent at once six medical officers to examine into the disorder; it also took wise, energetic and absolute measures to prevent it from ascending the river. These measures were effective for a while; but *commerce* broke down the barriers. An infected brig from Bakou, on the west coast of the Caspian, entered the Volga July 20, 1830, gave the disorder to Astrachan a second time, where 4,043 died. It spread to Saratof, August 12; Samara was infected September 3d, Kazan the 9th, and by the 12th a few cases appeared in Moscow.

19. We will not follow its other eastern route through Russia to Nijne-Novogorod, where it broke out August 27th, and thence spread through the centre of Russia. There is some uncertainty whence this

2*

irruption came. Some assert that it was due to the great annual fair of 100,000 merchants assembled to purchase the costly shawls and furs that came from Orenburgh—goods that had been infected and lay there since the previous year; others say that it might have been brought by travelers from Saratof, where it raged fifteen days before. September 3d it reached Kostroma, and many towns and villages on the route to Moscow.

20. By another route it crossed from the Volga, where this river approaches the Don, and descended this last named river to the Sea of Azof and the Black Sea coast in September, reaching Sebastopol in October, and broke out in Kischrif in December. Its ravages among the Cossacks of the Don were frightful; out of 54,000 cases 31,000 died—that is, out of seven patients six died!

21. About the end of September, 1830, cholera broke out at Moscow. On the 29th a military cordon was established around the city, whose office was to prevent egress and ingress. Strict quarantine was exacted of all—even the Emperor himself obeyed the law, and underwent eight days' quarantine on his road to Twer. Mortality was great here throughout October and November, but diminished, as usual, after a while, by the 18th December. The population was 250,000, of which 8,130 died, equal to one in

25, a much smaller number than was expected; but this diminution was due to the cordon, the strict military discipline and quarantine, and to the closure of infected houses and places. A triple line of posts preserved St. Petersburg. Orders were sent to the people to lay in provisions for at least one year, to meet the wants of seclusion, should that become necessary.

22. The disorder still existed in January, although a severe winter had frozen all travel and communication. Still, all the sanitary regulations were kept up until the end of April, in consequence of the experience gained at Kiew, where, the disorder having ceased in November, all restrictions were removed; but unfortunately so, for only fifteen days after the relaxation of all restriction it broke out again, on the 15th December.

23. The Russian army moving on Poland was the rapid means of carrying the pest westwardly. The troops of this army were drawn from the Ukraine and the provinces of Koursk, where cholera had raged the previous autumn. The violence of the plague did not diminish with its entry into a northern climate, for in April and May more than half of the attacked died, and this frightful proportion was exclusive of the military loss. Other troops coming from Bender and the shores of the Dnieper spread the pest among the people in February. With this army, cholera entered Poland at the southeast corner of the kingdom, on its march to the northwest, its destination being Warsaw. At the end of March it ravaged Lublin. By the 1st of April the hospitals of Siedlee became crowded with patients affected with cholera. On the 10th it broke out among the prisoners and the wounded brought into Praga, a suburb of Warsaw, on the opposite side of the Vistula. The same day it was discovered in the Polish army after the battle of Iganie, proving fatal to the conquerors by the infected spoils collected from the slain of the vanquished.

24. The Russian Government appointed an extraordinary committee in April, whose office was to discuss and decide whether it was necessary to have a general purification of the goods left in Moscow during the winter after the cessation of the plague there. On this committee there were several *merchants*, besides disinterested scientific men. A *majority* was to decide. It is easy to see that cupidity would overpower honesty; accordingly a *majority of votes* decided that purification was unnecessary. Acting on this dangerous advice, the Government issued a circular to the several European powers to justify its conduct in not purifying these goods before their export. The result soon became apparent, for the boats which carried these goods down the Dwina and its tributaries to Riga carried also the contagion, which broke out in Riga 25th May.

25. With the irruption of the fell demon there also broke out alarm. On the 3d June sixty vessels fled in haste from Riga, and forty-two had already passed the Sund. Four of these vessels were destined to England, and one that was infected entered the port of Montrose in Scotland, where cholera spread lightly in June, but of which no official notice was announced.

26. An official report declared that 306 cases and 94 deaths had occurred in Sunderland, by October 26th.

However, Mr. Hall, surgeon of H. M. ship Temeraire, an hospital ship in the Medway, says that between the 7th and 9th August he had thirty cases, persons of all ages, even to infants at the breast. Drafts from Portsmouth, Plymouth, etc., who arrived to join the men-of-war here, men in robust health, took the disease within the first and second day of their arrival. The infected ships were sent into quarantine at Stangate, ten or twelve miles distant, by which means the pest was prevented from spreading there and then.

To return to Sunderland, it is affirmed that cholera was brought into that town through clothing collected from the dead in the Baltic.

27. The cold winter of Northumberland afforded no obstacle to its propagation; even Edinburgh suffered in February, 1832.

28. The object of writing this book is to record the first entry of a new pestilence into a new country, to be substantiated by positive and precise data. But it was deemed proper to precede this event by a few words on its origin and migration towards the " New World." Therefore it is that we have given the foregoing rapid sketch of a pestilence that has spread its long arms to the four points of the compass, reaching from the place of its birth in the Sunderbunds to the Baltic, and thence into England— the *leap-off point* from the old to the new continent.

However, before touching on the history of its arrival into the towns and parishes on the St. Lawrence, and its subsequent spread in America, we must not omit a summary and exposé of the important and interesting opinions of writers on its origin or remote cause. This will constitute the second chapter, in which we shall have to repeat much of the itinerary already stated in this first chapter. After the fourth chapter, we shall trace it—lead, as it were, by the hand—across the ocean into the " New World."

CHAPTER II.

In this chapter on the remote cause will appear a succinct notice of opinion set forth by non-contagionists, and a history of its march from place to place, and from individual to individual, proving its contagious character.

29. The remote cause of all diseases, especially of pests, is the most important point to become acquainted with; for with this knowledge the practitioner will know what precautions ought to be adopted against the disorder and its spread. With this it is the duty of the magistrate to become thoroughly acquainted, in order to arrest its invasion of his country, and the spread of an infectious disease among his people, since he is appointed to watch over, **care** for, and protect them. **To** attain this knowledge, the widest range of research and discussion must be accorded to all men who undertake an examination into the subject. Without a knowledge of the remote cause of a pestilence, no well ordered measures can be employed to exclude the misery it

will carry into a country, or measures to mitigate its power for evil when once it prevails; without such knowledge, every precaution taken, every means and effort to control it, will amount to confusion, great errors, injurious restrictions, even oppression, while all the time the best and most anxious intentions are in full exercise.

30. In the early days of Indian cholera, practitioners appear to have vied against each other which should be the most ardent argumentator in support of the assertion that the pest was due to something else than contagion.

There always appears to be something peculiar in the medical mind which prompts the votaries of that obscure science to seek elsewhere for explanations than among the simple, every-day facts that strike the common sense of the vulgar. This has ever been the case on the occurrence of every known pest, and never yielded to common sense and observation until the imaginative, metaphysical and argumentative practitioner has been forced to abandon his fine-spun hypothesis.

Let us refer to a few of the pests that have afflicted mankind, and we shall find that the same contentious professional spirit prevailed in every one of them.

31. *Plague*, as it is emphatically called, is one example, which, until very lately, eminent partisan

men, but not the vulgar, attributed to climate, to terrestrial exhalations, filth, etc. Even such men as
Baron Desgenettes, a man of unquestioned science,
but also a man of crotchets, boldly contended that
plague was not contagious; and, to support his assertion, mixed freely among the French infected in
Egypt; even went so far as to inoculate himself with
the matter of the pestilential bubo; and because HE
escaped, cited his case as proof irrefragable of his
promulgated opinion. Several enthusiastic young
French doctors imitated the example set by their admired chief, inoculated themselves, and died. But
to the partisans of a theory the infection of the
young doctors was easily explained and denied, or
rather, ignored; the partisans, whose greatest merit
is adhesion to previous opinions, boldly declared
that the disease of these young enthusiasts was due
to the common unseen cause—an epidemic—and not
to the inoculation. Many medical writers appealed
to the example afforded by the conduct of the commander-in-chief of the army going among the infected, and escaping. It was gravely asserted that
he escaped because he had no FEAR of the disease.
It is well known that Bonaparte, like any other
great commander, would venture almost any thing to
preserve the confidence of his army, which, if lost, all
was lost. This story about *fear* was one of the

stalking-horses much paraded during the prevalence of cholera in Canada and in the United States.

32. Another example of medical contention is to be found in *Typhus.* This fever arises spontaneously in many places; and, when there is only *one* case, or only a few, here and there, it does not spread. This fact was the origin of the name *Typhoid;* but, said the originators of the new name, it is not typhus, because it does not spread—*morbus contagiosus non est!* In certain epidemics of the disease and period of its duration, there might, or not, be a cutaneous irruption of petechiæ, vibices, maculæ, etc. Each of these accidents to the disease sufficed to furnish to the imagination of would-be profound observers several new names to a well known and heretofore well described disorder, such names as spotted fever, enteric fever; to autopsists, the names gastro-enteric phlegmasia, ulceration of the agminated and solitary intestinal glands, mere accidents to the disease; and lately we have been favored with a newer name—cerebro-spinal-arachnitis! But when typhus spread incontestibly from person to person it was said to be *infectious*, but not *contagious*, a distinction that tickled the author into the belief that he was a profound observer and a learned etymologist.

33. *Small-pox*, another contagious disease, was

long ignored as such, and writers say that Sydenham was not aware of the fact. It was first noticed in Europe in 572, and in 580 the Bishop of
Avranches, in France, gave to it the name Variola,
a Latin word of his own coining. It desolated Gaul
and Italy, and in 580 Austregilde, wife of Gomtran,
King of Bourgoigne died of it. She was so incensed
that she forced a promise from the king that her two
physicians, Nicholas and Donat, should be sacrificed
on her grave as a punishment for not curing her.
The crime was perpetrated. But variola had long
been known in China. It was spread through Europe by the Saracens who overrun the country. In
Georgia the discovery of inoculation was made, and
necessarily spread to Constantinople, where the
famous Lady Montague learned it, and, returning to
England, prevailed on the Princess of Wales to set a
public example by having her children inoculated, in
1716. This met with violent opposition by the profession and the clergy. It took a long time to introduce it into France, where one of the first to undergo
it was the daughter of the famous Montcalm, in 1764.

34. A negro, from St. Domingo, brought variola
into the army of Cortez in Mexico, June, 1520,
where it annihilated whole villages and towns; the
dead were too numerous to be buried, and the stench
they created was so great that the houses were fired

and in them the dead burned. For a long time variola, like cholera, was denied being contagious.

35. Measles and scarlatina were also brought into Mexico about the same time as variola. All these diseases were hitherto unknown in America.

36. Syphilis belongs to the above category, but for reasons that may be appreciated is passed over. Nor need I refer to the black-death and the Sudor Anglicus, both of them limited pests.

The four pests mentioned are referred to simply to show that contagion, admitted from the first by the vulgar, has been generally denied for a while by the scientific; and is here introduced as a preparatory caution to the reader before he goes through the examples cited in this chapter on remote causes, wherein it will clearly appear that cholera never preceded, but always followed, the arrival of infected persons.

37. If we contemplate with unbiased minds the itinerary of cholera, in India and elsewhere, it will appear that the malady spread from individual to individual; that it nowhere broke out until communication had taken place with the affected; that it never, in a single instance, preceded the arrival of affected persons or things; that it often broke out almost immediately, and rarely later than thirty-six hours after the arrival of the infected; that it

rapidly attacked a fourth, more or less, of the inhabitants of a place previously healthy, and that at least one-third of the attacked died; that, in a limited population, as that of a small town or village, or a concentrated one, as on ship-board, it soon exhausted itself, and ceased to rage in two or three weeks, or less time; that when a large town was attacked, it would take from two to three months to select all the susceptible, and then cease.

From these data we are forced to believe that the remote cause is a poison generated in a patient, and emanated from his person. This theory is supported by the fact that seclusion invariably protected the secluded. We can only suppose it to be a poison which travels independently of wind, climate, and season—a poison given out by man, and carried by man to man. It is of no consequence what place or person afforded the starting-point—it is individual in every patient and place; its cause is ONE and identical everywhere. Tropical heat made no difference in its violence or duration from that of a Russian winter; the atmosphere neither carried nor opposed its transmission, since it traversed the broad ocean against the trade-winds. Famine and peculiar food made no difference, nor did races of men, since it broke out with equal severity in the palace of the prince, the mansion of the nabob and the rich merchant, and the hovel of the peasant and the poor.

38. Why and how this greatest of all known pests originated and spread with appalling rapidity in all directions from the place of its birth, regardless of topography and seasons, has been largely commented upon, as might be well expected. For several years after its irruption in 1817, the majority of opinions for a while seem to attribute its propagation to occult causes outside of its victims; yet it is curious to notice that many "medical boards," that were appointed to report on the "disease," manifested in the reports a strong bias in favor of the opinion that it was propagated by the persons affected—curious, we say, to find the same names attached to the *official reports*, as are found outside of the reports on the title-page of numerous monographs written to promulgate an opposite opinion, that is, that the disorder is not communicated—is not contagious; and these opinions are supported by "facts" and much metaphysical reasoning. It would seem that the officer charged with a responsible duty towards his government felt shy in making a hazardous and dangerous declaration, but who, left to a flighty imagination, felt himself now free from responsibility in his private writings.

39. To deal fairly with the non-contagion theo-

rists, it is only justice to say that the faculty of
Moscow, who knew the disorder only from printed
accounts, were all contagionists; but shortly after
they became personally observant of the plague, in
1830, soon changed from contagionists to *non*-con-
tagionists. This change of opinion they openly proclaimed, and supported it by writings in which they
justified their new opinion, by stating that medical
officers, students, nurses, etc., escaped, although much
exposed. To superficial observers these " facts"
seemed irrefragable, and they are even now appealed
to by medical men given to polemics. However, we
must again do justice to the practitioners in Moscow,
by recording that in a few weeks they once more
changed opinion, seeing that the " facts" they relied
upon had been erroneously observed, and that the supposed exempt were soon numbered among the victims. It now became manifest that what appeared to
be a fact yesterday, ceased to remain a truth after
more extensive inquiry.

40. This alleged impunity of exposed persons has
been from the beginning and throughout the reign of
cholera the one—the only—argument against the
idea of contagion; and was, and has been, contested
with a partisan mode of argumentation which has
brought into strong relief the mental capacity of
some of the writers for metaphysical reasoning, bril-

liant as well as confused ideas, and tenacity of purpose.

41. Among the assigned causes that originated and perpetuated the malady was *the atmosphere*—that convenient receptacle for the location of all kinds of medical ignorance. Chemical admixtures, physical admixtures, even metaphysical admixtures, were declared to exist therein, and were proved to the satisfaction of each author, however widely they might differ one from the other. For at least fifteen years the medical press teemed with these lucubrations, caught up by the newspaper press, and made clear " to the meanest capacity." But a necessary condition for the atmospheric cause to operate was some anchorage ground in the patient—he must be " predisposed"—he must have taken cold—his system must be in an " abnormal" state—his digestion must be disturbed, etc.

42. Numerous equally absurd statements were put forth: atmospherical and terrestrial *commotions* were accompaniments, or rather causes, of cholera; rapid change of temperature from cold to heat ; barometrical fluctuations, unwonted meteors, whirlwinds, waterspouts, fogs, volcanoes and earthquakes, overflow of rivers, geological changes that are continuously and silently going on. Influence of comets (the one seen in November, 1831 or '2), extraordinary *Lumen Zodiacale*, seen September 29th, 1829.

43. Remarkable *aurora borealis*, 7th January, 1830, and another in August; a new island upheaved in the Mediterranean; clouds of hitherto unknown green insects in the upper strata of the atmosphere, brought down on the tails of kites at Astrachan, reported by a learned naturalist (nothing like leather to fortify a city, said the cobbler). But Dr. Forster says: "I give little credit to this statement as a cause; for I ascended, on the 30th April last, to a height of 6,000 feet in a balloon, and found none of these pestilential flies which some experimentalists pretend to have come down on the tails of kites." A tornado in Barbadoes; earthquakes in Venice and Parma. All the authors of the foregoing inanities seem to forget that cholera ravaged the world at least thirteen years before the occurrence of the foregoing statements. One of the casuists boldly declares: "I am convinced that if a premium were offered for the importation of cholera or yellow fever, by means of dress or bed-clothes in which patients had died, the thing would be found impossible. No mischief would arise, unless the importers of the bed-clothes could import the *peculiar* constitution of the atmosphere and the earth under the atmosphere where these maladies originate." What a waste of time and money

3

to have taught such a man to write! Another phi-
losopher says: " Specific local atmospheres produced
by irregular streams of specific atoms from the interior
of the earth.   Perturbated electrical and magnetical
states of the air and earth have been the hobby ridden
by many a daring medical Mazeppa."

44. Another cause assigned was great humidity of
the air in some places, over swamps and deltas.
This supposed cause was announced by practitioners
who lived in humid localities, and who heeded not
that cholera reigned equally supreme in arid countries
—such as Arabia, where rain scarcely ever, or never,
falls, the land parched to a great depth, and a spring
of water so great a rarity as to be sought for at a
distance.

45. Others attributed it to vegetable and other
putrefaction exhaled from low places; while it is well
known that such putrefaction did not and could not
be a cause, since aridity was the rule in some countries
where cholera prevailed.   Another statement, much
in vogue for a time and widely disseminated by journals and common newspapers, was that cholera followed the "great lines of commerce."   We are left
to suppose that it followed these lines as a mere
traveling companion—never preceding commerce or
lagging behind.   The way the wind blew did not
disturb the traveler, for he, cholera, traveled in the

" wind's eye" rather than abandon his companion, commerce; and all this time the atmosphere was the abode of the pest that moved in a contrary direction. I well remember with what dogmatic assurance a merchant, newspaper editor, or other pretender to knowledge, would cut all statements short by throwing the hackneyed phrase in your face, " it follows the great lines of commerce "—reply to that, if you can !

46. While being carried, according to some rea-soners, by the atmosphere from place to place, it would descend to attack victims in one spot and spare an adjoining place, over both of which this atmosphere spread equally. Cholera would attack one side of a river, the road side, and neglect the other; attack the inmates of one house and not those of an adjoining one; attack some portions of a town and not others, while the same atmosphere enveloped all sides and over all equally.

The foregoing statements are only a few of the many put forth and bandied about for several years, to the disgrace of the human understanding, and should not now be noticed were it not that there still linger some of those credulous or obstinate beings who are an annoyance to the profession.

To descend from the extravagant and wild flights of imagination like the foregoing, let us notice some of the more reasonable opinions at first put forth regarding the remote cause, or causes, of this dire visitation upon humanity.

47. *Moisture* a remote cause. Let us examine this alleged cause. It was natural enough for practitioners living in the extensive flat, low and swampy country of the Bengalese territory—a country intersected in all directions by numerous dikes, rivulets and rivers, tributaries of the great Ganges, which opens into the sea through a nearly stagnant delta of many branches, called the Sunderbunds—to attribute its malignant character and origin to the nature of such a country, the capital of which, Jessore, low, flat and wet, like the country around for many miles. Besides this disadvantage, it is a remarkably ill kept and filthy city, containing a dense population.

It is here that cholera is alleged to have first put on the malignant and spreading form. To these sanitary defects add the hot season, which commences in March under a south wind, the temperature rising from 70° to 90°. The rainy season commences in June and continues until October, before which time the Ganges reaches its highest point, and inundates the country to a vast extent. The cool season, which is never cold, commences in November and lasts until February. During this time the sky is clear, the air cool and bracing, the thermometer ranging from 45°

to 75°, with winds mostly from the northwest. But, although it does not rain in this season, copious dews prevail at night, owing to the great daily evaporation and nocturnal precipitation.

48. It is not to be wondered that the early writers on this pestilence should have attributed its *new* and malignant character to such moisture, and to a filthy city. Moreover, in this year, 1817, the amount of rain that fell was quite one-third more than in former years. The inundations cut off communications, and thus increased the scarcity of provisions and deteriorated the quality of what remained.

The pestilence soon spread to Calcutta in apalling force, among the poor first, who are improvident and intermingle freely and without discrimination. It soon after attacked the better classes with equal virulence, but in less numbers, because they are less numerous and more guarded in their intercourse. By September the villages and towns for several miles around were all infected.

49. Now, as if to baffle and joke with medical opinions and statements, the following indisputable facts occurred: While cholera traveled northwardly from Calcutta in this same year, 1817, between Agra and Delhi there are many villages situated in low grounds, but these *all* escaped, while Cawnpore, Meerhut, Agra and Delhi, all high and dry grounds,

suffered *severely*. The fortress of Jaragurth, 1,000
feet above the plain, suffered greatly, and this while
the inhabitants of towns at the foot of the mountain
escaped. These high and great towns were rendez-
vous of trade, the government commissariat and
troops.

50. If only low places, like the deltas of the Ganges, the Euphrates and the Nile, etc., had been attacked, *non*-contagionists might say, with a strong
show of probability, that cholera was due to marshy
miasm.

We shall now proceed to show that low grounds,
swamps and deltas were no more obnoxious localities
than the high, dry and healthy parts of the world.

51. *Dry grounds and high countries.* The high
mountains which separate Hindostan from Nepaul
were invaded by the pest, as well as the elevated
table lands of Nepaul, Patum and Rhatgoun, 4,000
feet above the sea. Catmandou, on the lower range
of the Himmalaya, at a height equal to the Pyrennes,
suffered greatly. Surely, this is land high enough
and cold enough to contrast with the steaming swamps
of the Sunderbunds and that hot climate.

52. The table land of Mysore, Darwar, Belgaum,
Bengalore, all very high, over 4,000 feet above the
sea, were attacked with equal fury and destruction to
lower and hotter places.

Cholera crossed the Ghaut mountains that separate the east of India from the coast of Malabar, a mountain range as high as the Carpathians or the Pyrenees in Europe. It invaded the town of Candy, situated at a great height, in the island of Ceylon, as early as December, 1818, having commenced at the coast and spread to the interior.

It ravaged the Pachalicks of Syria, and crept up the flanks and over the top of Mount Lebanon, in 1823; attacked the town of Tiberias in the winter of 1824; ravaged the Caucasus, the flanks of Ararat, the Himmalayas, where its violence and destruction were equal to what happened in low and hot places. Nor did the valley of the Jordan escape, a country 1,400 feet above the sea.

53. *Cold.* Besides the elevated countries just named, which are naturally of low temperature, special mention may be made of the invasion of cholera in the north of Europe, where the country is flat, but not wet in the winter time, since all surface water is frozen over. The pest reached Moscow in October, and persisted there until its decline at the end of December, 1830, a period of nearly three months, quite as long as it lasted in any invaded hot place—apparently the allotted period of its duration and power in every place it has invaded. From Moscow it traveled north to Yarosof, and, turning to the west, reached

Rybinsk, sixty leagues north of Moscow, on the road to St. Petersburgh, on the 19th March, 1831. This calendar of its march shows that the severe frost of an intensely cold country did not arrest it or alter the quality of the pestilence.

It is now evident that neither low nor high localities, nor warm nor cold regions made any, the slightest difference, in the spread of cholera, its severity or its nature.

54. *Geological formation.* Cholera swept the surface of the Arabian deserts, a country so dry that showers are unknown, and where a spring of water is the object of a journey of 300 miles. It desolated the dry calcareous plains of Arabia with the same fury as the deltas of the Ganges, the Euphrates and the Nile, the swamps of the Volga and Dnieper; traversed with equal ease the sandy plains of the Yemen, the basaltic declivities of Mauritius and Bourbon.

55. *Communication from place to place, and from person to person.* In this section we shall notice only the best authenticated statements, without following, *seriatim*, the propagation from Jessore, whence cholera ascended the Ganges, attacking place after place, each having previously given passage to infected persons, until it reached the healthy banks of the Sinde, attacking first the numerous followers of the army of the Marquis of Hastings on the 6th of November. By the 15th, in nine days, it spread throughout the camp, attacking all ages and sexes. In five days' time 5,000 deaths took place— one thousand a day! In this short space of time it had reached its acme; then commenced to decline until the 23d, barely two weeks, when few cases occurred, and the deaths that now happened were those cases that had lingered on for several days.

56. The Marquis, now that the storm was over, commenced to move his camp to the southeast, and did not complete the march until the 19th December, a period of 26 days. During this march, very few cases broke out, and none after the 8th of December. On the foregoing, Mr. Kennedy lucidly remarks:

" A superficial acquaintance with the progress of cholera might lead to the supposition that its decline was connected with the change of locality prescribed by the commander-in-chief; but the history of the disease shows, on the contrary, that it had *run its course* of infection before the army quitted the banks of the Sinde, and the only benefit derived from the change of air was the assistance it afforded in the recovery of the multitude of people who lingered under the effects of an old attack. To ascribe the cessation of the pestilence to any virtue in the soil or atmosphere of the encampment by the Betwah would be little less than absurd; for the troops did

3*

not arrive there until the 19th December, and a new case of cholera had not been observed during the eleven days which immediately preceded their arrival. If additional evidence were really necessary to settle this point, it would be amply furnished in *the law of increase and decline* appertaining to cholera. This law informs us that when the malady has been developed in a camp it will continue its ravages through the period of one month or so, independent of locality—whether the infected camp be removed from a moist to a dry district, or from a low to an elevated station—and, at the expiration of the month, that the disease will die away spontaneously."

Staff-Surgeon Connell reported, Secunderabad, May 20th, 1819: " A detachment of Europeans, under Major Wahab, arrived here with the cholera among them. The disease first attacked these troops at the Kistnah, after exposure to a great storm, and it continued with them to this place, although all the villages in their route were entirely free from the disorder. During the march sixty individuals perished, of whom eight were Europeans. On its arrival here, the detachment encamped about two hundred yards in front of our artillery lines. In this new situation, three Europeans and a number of natives died. Up to this time no case had occurred in the encampment. The Europeans, however, of Major Wahab's detachment mingled with our party of artillery, and in the course of four or five days the disease began among the latter. * * * * *

* * My sub-assistant, Mr. Hoskins, who was constantly with the sick, took the disease and died in twenty-four hours. Another sub-assistant, Mr. Steven, and Mr. Assistant-Surgeon McDougal were attacked. The disease traveled to the bazaars and carried off many of the natives. The men of the Thirtieth regiment, who were in barracks, half a mile to the right, completely escaped.

" I beg to add that Mr. Jones, Surgeon of the 6th Light Cavalry, has just arrived from Kistnah, by the same route as Major Wahab. Mr. Jones states that he found the cholera prevailing in every village, having commenced soon *after* the passage of Major Wahab's men. The inhabitants said they had got it from that detachment."

57. After the irruption of 1824, Madras remained free and healthy until March, 1827, three whole years. But in July it suddenly broke out with violence at Jaulna, Hyderabad, etc.; during this irruption, the governor of Madras, Sir Thomas Munro, died. He sat down to breakfast at Pullycondah, a village twenty miles from Gooty. While at breakfast he was taken with the first symptoms, and said to his secretary, who was beside him, "*Get away; I am an infected man*," and he died *ten* hours after.

58. The pest reached Suedie, Mediterranean coast of Africa. But it was not very destructive here. However, Mr. Baker, British consul, reported that he had about twenty natives at work harvesting for him, all robust and healthy men. On the 9th July, at noon, when all appeared well, one was suddenly attacked, and shortly after others, with vomiting and cramps. Some died in *three* hours, others at sundown, and all were dead before daylight. At this time strong westerly winds prevailed, which extended one hundred miles into the country, but did not drive the pest back to the east, whence it came. "It had traveled in the *wind's eye*."

59. A detachment of troops brought a state prisoner from Panwell, where cholera raged, eighteen miles from Bombay, to Tannah in the island of Salsette, where it at once broke out. Thence it spread from village to village; but several villages, which had no intercourse with the infected places, escaped; however, after several months' immunity, they, too, were attacked through a renewal of intercourse.

60. Cholera broke out in the Thirty-fourth regiment on its march from Bellary to Bengalore. All the villages on the route were healthy; but immediately after all the villages it passed through became

affected. A native soldier, traveling from Bengalore to Nundedroog, both places then healthy, passed through the Thirty-fourth, was attacked and died. Nagpore suffered in May. A brigade in perfect health arrived at Nagpore, now an infected place, on the 4th; on the 6th the malady broke out among the men with great mortality. The English army, in the best of health, encamped at Terayt, and received a detachment which had been attacked at the passage of the Jumna. Immediately after the junction the malady appeared in camp. A company which had lost some men on its march arrived at Trichinopoly, then quite healthy; the pest immediately broke out among the people and spread to the environs. The Fifteenth native regiment, affected with cholera, marched on Gooty; the villages through which it passed immediately became infected and desolated by this scourge, from which the inhabitants had ever before been exempt.

61. Muscat, a seaport at the extremity of the Arabian peninsula, much frequented by traffic from Bombay, where cholera prevailed, became infected in July, 1821. About the same time the islands of Ormus and Kishmé, at the mouth of the Persian Gulf, became infected by trading vessels. Bender Abassi sunk beneath its rage, and in a short time *one-sixth* of the inhabitants perished; the bazaars were closed, the

houses abandoned, unburied dead lay in heaps in the streets, the surviving population having fled.

62. *Poland and clothing.* After the battle of 31st March, 1831, in which the Poles were victorious, they encamped in a morass, and remained there eight days. April 10th a part of this division fought before Siedlce with the corps of Count Phalen, which was infected with cholera. On the 13th, while passing through Kuflew, a report was received that *six* men had died of cholera after a few hours' illness. These men formed part of the first brigade, which had captured colors and several prisoners. Arriving near Minsk the cases became more numerous, and on the 15th fifty had died. The majority of the patients had *on them clothing* taken from the *sick enemy.* The second brigade, which had not been engaged at Siedlee, had no cholera for a length of time. Later, another division, in a state of perfect health, encamped near Kuflew, on the ground where the Russians had just been defeated; several bodies still lay exposed, others only half buried. Here cholera broke out anew among the Poles. Towards the end of May the division had a serious engagement at Tycocin, and cholera broke out among the Poles for a *third* time.

63. About 200 tailors took as assistants from among the Russian prisoners, some of them to work.

Cholera *immediately* broke out among the former. Russian prisoners were dispersed over several villages and towns, and wherever they went cholera coincided with their arrival.

64. The last great battle fought by Count Diebitsch was on the 26th of May, and the result was equivocal. Diebitsch took ill success to heart, was seized with cholera, at Pultusk, on the 10th of June, and in a few hours died. Also, on the 27th of the same month, the Grand Duke Constantine died, at Witepsk, of cholera.

65. Nijne-Novgorod, in the centre of Russia, was attacked 27th of August, immediately after the great annual fair, when 100,000 merchants had assembled to purchase the shawls and furs that came from Orenburg, where these goods were stored over the previous winter, an infected place. But other writers say that cholera might have been brought by travelers from Saratof, where cholera raged fifteen days before the fair. Both accounts are likely to be true.

66. *Enclosed places escaped.* The great city of Gourruckpoor and environs lost 30,000; but *not one* of the prisoners confined in the jails of the city was attacked. The frigate Topaz, from Calcutta, anchored in the port of St. Louis, Mauritius, November, 1819, and had lost men and officers on the passage, but all were well on arrival at St. Louis. The haughty captain refused to obey the quarantine; he and his men went on shore, and soon after the pest broke out and spread rapidly—many were attacked while walking the streets, and died there almost as soon as attacked, so severe were the cases; 10,000 are said to have perished; but the governor reported that the deaths did not exceed 7,000 or 8,000. Certain habitations that interdicted all communication *completely* escaped.

67. The neighboring island, Bourbon, in fearful proximity with Mauritius, became alarmed, and the governor, Baron Melius, at once established a lazaretto, lest it might be needed, and interdicted all communication from outside. This was effectual for a while; all escaped, until a smuggling vessel, the Pic-Var, clandestinely landed some runaway negroes, December 14th, at St. Denis, where the pest at once broke out. The governor immediately established a military cordon around the city, suffered no one to go out or go in. The attacked were sent to the lazaretto, which was also surrounded by military. By these means the country was saved from the spread of the pestilence. The contrast between the two islands was striking: Mauritius had 20,000 cases, half of which died ; Bourbon had only 259 cases, and 178 deaths.

Dr. Labrousse traced the infection step by step from the place where the Pic-Var landed the smuggled negroes into Bourbon first into two cottages, in one of which six negroes were attacked, and two in the other. The inhabitants, frightened, interdicted these places, and here the pest was arrested. At one of these houses, a negresse, Mamédé, wife of a fisherman, infected her husband, and he died. She went to the residence of her master, a mile off; the next day she was attacked, and communicated the pest to a servant slave. The jail prisoners employed to convey the cases to the lazeretto died in the service, and only two nurses escaped. In the hospital the pest attacked not only the servants, but *patients* laboring already under *other* diseases. The doctor exclaims: "If this be not contagion, how comes it that a few sentinels effectually opposed its efforts to cross the military cordon?"

68. When cholera ascended the Volga the Moravian inhabitants of Serepta shut the gates and harbors, suspended all communications with persons and things from outside, and thus excluded the pest and escaped.

69. In Moscow cholera was arrested by sequestering houses where it existed.

70. In 1822 Mons. de Lesseps, French consul at Aleppo, fearful of the approach of cholera, shut himself up and his people, to the number of 200 persons, in a large garden not far from the city. The garden was surrounded by a wide fosse and a stone wall, with a gate at each end, which he kept guarded, and prevented egress and ingress. Not one person in this enclosed colony took the malady, while 4,000 died in the city in eighteen days.

71. In November, 1819, cholera attacked Sankerre-droog, and reigned there until the 14th December. But the mountaineers forbid any communication through the passes into the valley, by which precaution the pest was excluded and the inhabitants escaped its tyranny.

72. The Pacha of Egypt adopted severe measures in July, 1824, and by them the disorder that prevailed in Syria was prevented from entering the valley of the Nile. This example, with that of the Shah of Persia at Teheran, proves, among many others, that the contagion can be arrested by exclusion. But at last, in 1830, news reached the pacha that a column of fugitive pilgrims had already passed Suez, and a second column had proceeded beyond Coffeir, the only points of communication with his dominions. Instant orders were sent to enforce quarantine in these places; but the early travelers had already passed, and by the 13th July 4,000 had arrived at Cairo. Cholera broke out first at Suez,

then at Coffeir, shortly after the arrival of the fugitives. Cairo next suffered. The physicians and apothecaries died; the board of health abandoned their useless efforts; people fled the city, and those who remained shut themselves up, and *all these* escaped.

73. Caravans, infected, traversing from the south to the north, when nearing Teheran, the residence of the shah, alarmed him. He at once issued orders that the caravan should not approach the town. In consequence of this mandate, the caravan made a long detour, infecting as they went the places they passed through, but the city escaped by this precaution.

74. *Ships.* The frigate Leander anchored, August, 1820, in the port of Pondichery, where cholera raged. She soon became infected and put to sea to escape. On her voyage she lost *ten* men and *two* officers, before she reached the port of Trinchimalee, on the 11th July, where no disease existed. Soon after her arrival cholera broke out among the persons and officials that communicated with the ship. The surgeon of the Marine Hospital was the first victim. It soon spread through and through the island, previously healthy. By coasting vessels it was carried across the strait into Colombo, January 10th, 1819.

75. A brig carried it to Astrachan, and by boats it reached the Sea of Azof, Sebastopol, etc. All the

fishing villages on the Arabian and Persian shores of the gulf escaped, while the great ship harbors and entrepots of commerce became immediately infected after the arrival of vessels from Bombay.

76. The frigate Abercrombie left Bombay August 10th, 1828, a diseased place, in splendid condition and fine weather. The same day cholera broke out on board. On the 12th the pest was general throughout the ship. Many men died in *six* hours.

77. At Collapore the following occurrence is said to have taken place, no doubt much exaggerated. *Sixty* persons embarked to cross to the opposite shore, and only *three* were able to disembark, all the rest perished.

78. The French frigates Cybèle and Cléopatre— the first put into Malacca and took in supplies for the crew on the 14th. On the 18th she put to sea. The fourth day out the disorder broke out on board, and resisted medical treatment. On the 22d January, 1822, the Cléopatre anchored in the roads of Manilla, where cholera existed fearfully. On the 30th the disorder broke out on board so severely that the captain departed for Macao, and in eight days after no new cases occurred.

79. *Extreme cleanliness and hygiene* made no difference as a preventive or a mitigative of the pestilence, notwithstanding the clamor of ignorant busy-bodies

and meddlesome newspapers. Attend to these facts.
In September, 1821, Shiraz became infected by commerce from Bender Abassi. The harem of the prince-
royal of Persia was almost the first arena of its fury.
One of the first to succumb was one of his wives, besides
several of the beauties in the seraglio, and some of
his children. The palace of an Eastern prince is the
embodiment of all the luxury and refinement that is
in the power of man to acquire; grandeur, elegance
and riches are the necessary accompaniments of the
establishment; want, in the slightest degree, is unknown, and suffering, if any, can only rarely be felt,
and that by the few ladies therein whose organization
or moral sense may possibly here and there revolt at
their servitude. Religion also comes in as a portion
of the discipline, hygiene, and extravagance of the
place; and, in this instance, is principally confined to
frequent ablutions, extreme cleanliness, and a few
easy prayers. The dresses are of the most costly and
exquisite qualities, changed several times a day. To
these habits all the attendants are scrupulously observant. If here, where cleanlines and the most perfect system of hygiene that can be sought for exists,
and still cholera invades such an abode, and rages as
fearfully as among the most abject, poverty stricken
and uncleanly people, why rail out against dirty
streets *as a cause*, and neglect more important and

serious matters? The palaces and families of the Indian princes, as at Delhi, etc., etc., escaped no more than the villages, unless it was due to *exclusion*.

80. *Religion as a prophylactic and a cure.* Priests make a religion, indoctrinate the ignorant into a belief that they are the elect and preferred of the Deity to all outside of their fanaticism. This is the moral history of man, and flourishes in India.

In 1818 the usual multitude of believers assembled at the festival of Jatra, in Punderpore. In a few days three thousand fell victims to both their faith and the cholera. These pilgrims, terrified at the pest, fled to their homes, in spite of their faith and the power of the priests; and wherever they went, there cholera broke out.

81. Eastern superstition (religion of priests), in its severe and unrelenting precepts, scoffed at the power of cholera. The votaries, in obedience to the precepts, assembled one hundred and twenty thousand to honor the shrine of Juggernaut—under the wheels of the car of the idol the most faithful annually took pleasure to prostrate themselves and be crushed to death—were waiting for the awful presence; but before the priests could wheel it out of the temple and commence the sacred ceremonies, cholera, in contempt of the omnipotent symbol, broke out among the multitude, of which many thousands perished. The idol

remained in the temple incarcerated for several years—perhaps dared not to face the pestilence! The votaries fled to their homes, undefended by an ungrateful deity, and spread the infection in their homes.

82. When in June, 1827, the pest menaced Lahore, the people sought to arrest it by acts of devotion and charity. But Rungheet Sing, the Maraja, preferred to absent himself. He crossed the river Rari and encamped by the Kuttel-Khan gardens. By this course he preserved himself and his followers.

83. In 1820 the King of Siam, alarmed by the irruption of cholera in his capital, convoked his subjects on the sea-shore, there, by religious ceremonies, to anathematize the disease. The result was appalling: seven thousand persons perished on the spot, and, with the arrival of fugitives into all the districts of the kingdom, cholera also appeared among the inhabitants. Forty thousand died in the capital, Bankok. Cholera treated the religion of the priests and the faithful with utter contempt.

84. In the latter days of Chawl and the first of Zilcade (second week in May, 1831), the pilgrims arrived from Persia, the Indies, the Yemen, and other countries, suffering from the pest at the time of their departure. Crowded together, at Mecca, under every circumstance which could favor the propagation of the malady—exposed to a broiling tempera-

ture, wallowing in the putrefying heaps of blood and offal of victims sacrificed at the feast Coram Bairam —in four days 20,000 individuals perished.*

Many more instances might be cited, but the above are sufficiently disgusting.

85. *Range of Infection.* In no case has the range of infection exceeded a few yards' distance, especially when the number of attacked is few, or the place is airy or windy, which blows the poison away into a dilution insufficiently concentrated to poison the un-affected. Cholera commenced in the eastern wing of the barracks, and proceeded therein westwardly through eight companies of the Fourteenth regiment; but it suddenly stopped at the ninth company, and the tenth or light infantry company escaped with a

---

* *Wallowing in putrefying heaps of blood and offal.* This remark I suffered to be transcribed, that I might not alter the statement quoted—not that such putrefaction could engender or make cholera worse than if such filth did not exist. It has been abun-dantly proved by Bancroft, Parent du Chatelet, in their report to the Paris Council of Public Health, that workers in putrefying animal matters did not suffer from their disgusting trade. The butchers, glue-makers, catgut-string makers, knackers, anatomi-cal dissecting-rooms, etc., were not injurious. Orfila stated that the exhumed bodies, in a putrid state, kept several days under judicial examination, etc., never once disturbed the health of the operators. But a *popular* cry has been raised against dirt; of course, let it be suppressed, but not under a false accusation.

few cases only. There are many such events mentioned as inexplicable; but it soon became known that all the time a strong westerly wind blew through the barrack, which was open and well ventilated at the entrance end, to which end cholera did not reach; besides, a few yards' distance of separation is abundantly proved to be a sufficient guaranty when communications are excluded, or at least well regulated.

86. *Length of time from exposure to the break-out of the infection* is variable within a certain limit. It has seldom poisoned its victim in less than twelve hours after exposure; most commonly twenty-four to forty-eight hours, and rarely to the fourth day; but in the case of the French frigate Cybèle (§ 78), four to five days intervened between the exposure and the irruption.

87. *Duration* of the pestilence in each place. In a large and populous city, then, it would last from *six* to *ten* weeks, rarely to *three months;* among small communities and villages, it will have selected all its victims in a space of *ten* to *fifteen* days. This is easily explained ; for the free communication of the inhabitants of a small place, all known to each other, they soon have effected their intercourse with each other, while a much longer time must elapse in a large community before all can have com-

4

mingled who are susceptible of receiving the poison.

88. This is particularly remarkable on ship-board —a small, compact population. A ship puts to sea, apparently all well; in twenty-four hours, or a little over, one or two are attacked; by the third day nearly all the susceptible are down with the malady, and in a week more scarcely a new case happens. The dead are committed to the sea, and all hands remain well for the remainder of the voyage, from four to six or eight weeks. But what happens after entering port? In twenty-four to forty-eight hours after, persons having intercourse with the passengers or vessel are attacked—even passengers who had escaped the first onset of the malady, now that they get into a *new* and healthy atmosphere, are attacked and die. This fact has been frequently noticed in Canada, where persons, after remaining many days in an infected city, have ventured to travel away and were attacked on their route, in the new atmosphere, and died. These instances are numerous. Nor is this peculiarity due to cholera alone. In the early transit of passengers across the Isthmuses of Panama and Nicaragua to California, many escaped the malignant ague of that climate, who, on arrival at San Francisco, a healthy place, where ague is unknown unless imported, in a few days are down with

ague; even those who have suffered on the route and recover were subject to a new attack in the *new* atmosphere.

89. *Conquest arrested.* The victorious Abbas Mirza, fighting against the army of the Sultan, drove the Turkish army before him, battle after battle, until both armies reached Erzeroum, into which fortress the Turks betook themselves for safety, repose and resistance. But here a new and unconquerable foe appeared in the field. Cholera attacked the lines of Mirza, and turned his victorious troops and banners into impotence. His lines were smitten by an unseen enemy; his soldiers became frightened, threw down their arms, and fled in utter disorder—a real defeat!

The same thing would have happened to the army of the Marquis of Hastings, had not a most perfect discipline rescued it. As it was, the British columns in India have been retarded if not arrested in their operations more than once. The same thing happened to the Poles before Warsaw, and snatched victory from them.

90. *Drunkards.* Although the Hindoos and Mussulmans are a temperate people, exceptions are met with, especially in the large towns, where drunkards are sufficiently numerous to attract attention. It was noticed that, after cholera had swept a city, the proportion of known drunkards that escaped all attack

exceeded that of the well conducted people. In Can -
ada, where the free use of liquor is common and
cheap, for a while it was proclaimed generally, especially from the pulpit and by magistrates, that the
drunkards were carried off.

I denied this at the time, but was remonstrated
with on the ground that, even if I were correct, it
was of great moral importance that the statement
should go abroad. But in three months' time, the
pestilence ended, all the old notorious drunkards re-
appeared in public as heretofore.

91. *Bad crops and bad grain, as rice.* Dr. Tyt-
ler endeavored to show that bad rice produced cholera; but this was at a time when cholera was raging,
and no doubt many poor persons ate of bad rice.
However, it is well known that after even a good
meal of sound food, some unknown thing will cause
an arrest or vitiation of digestion, followed by spon-
taneous or sporadic cholera, resembling, in many of
its symptoms, Asiatic cholera; but all those reported
cases are defective in many points as regards cholera.

CHAPTER III.

92. *Cholera morbus* and cholera spasmodica, or Asiatic cholera, are two distinct states of the body. The first is a disease, the second is not. On account of this difference it is proper to mention at once the origin or etymology of the name. Cholera morbus has been in use since the time of Hippocrates, and is supposed by Celsus to be derived from χολὴ, bile, and ῾ρεω, to flow—literally, bile flux. It is so described by all systematic writers, who appear to copy from one another; all of them define the disease: *bilious* vomiting and purging, gastric pains, cramps in some cases, and prostration. It will not do for me to deny the accuracy of all authors since a thousand years past; but I am at liberty to say what I know, and to describe the cholera morbus I was called on to attend long before and since Asiatic cholera reached Canada.

93. The cases of cholera morbus, now often called *common* cholera, that have come under my notice, were *un*accompanied by bilious matter in the discharges. At first, when I was quite young in the profession, I turned to authors in hopes of finding some exceptional forms of the disease in which bile was not discharged; but I found all the books exact transcripts of their predecessors, and was forced to come to the conclusion that I either mistook the disease, or had met with one for which I could find no description. From year to year, until Asiatic cholera reached Canada, I was called to attend these nondescript cases, and often since. It is now time for me to describe the " common cholera" I have in view, and as *I* saw it. The cases invariably occurred to persons in previous perfect health, adults of both sexes; scarcely ever in the winter months, but might be expected in any warm summer night, more so when a thunder shower supervened to suddenly alter and cool the atmosphere. An individual, quite well, might eat a hearty supper with good taste and appetite, go to bed well, fall into a sound sleep of two or more hours' duration, and be awakened after midnight and before three in the morning with an uneasy feeling, nausea, quickly followed by vomiting and purging. The first matter thrown up was the supper, apparently *unaltered*, after this there was a continuous watery vomiting; the first stool or two consisted of the accumulated feculent matter, after that the intestinal discharges were also watery, without odor and without colicky pains. The discharge from the stomach was also without odor, or taste to the patient. In none of these discharges, either upward or downward, was there any *bile*, notwithstanding the convulsive vomiting which might be expected by its compression of the viscera, liver and gall blader, would be likely to emulge the bile into the stomach and intestines. In some cases cramps came on, the skin grew moist, and, in protracted cases, cold; in severe and protracted cases the voice failed a little; sanguineous stasis became apparent in the skin, but not to a degree amounting to the cyanosis seen in Asiatic cholera. The most prominent symptom was vomiting, the stomach being so sensitive as to be unable to endure more than a tablespoonful of fluid. It was this instability that made the difficulty to treat the case, for medicine in the form of draught was too voluminous to be retained long enough to calm the fretful state of the organ. On this account it was that the patient was directed to swallow, as best he could, a pill of one grain of pure opium, without any vehicle. If this could be retained for half an hour, it was sufficient to calm the stomach, arrest the discharges and cramps, if those happened to exist. It must be noticed that the patient is always very restless and tosses about the bed; so sure as this movement is permitted or indulged in, the opium pill will not remain on the stomach—vomiting will continue; the most perfect quietude, even of speech, must be persevered in. One pill will generally arrest the great disturbance, and the patient will be well the next day, excepting the debility and the soreness of the abdominal muscles. When the case has been permitted to go on for a few hours, prostration becomes too great to be rallied from, the cold surface continues, sometimes with slight lividity of the skin and nails; the pulse becomes imperceptible at the wrist, the heart acts feebly; consciousness remains clear from the first, at last becomes obtuse; restlessness ceases, prostration increases, and the patient dies imperceptibly, without a pang or struggle. In some few of these severe, unarrested cases, the colliquation will cease, giving hope of recovery. The patient does so to some extent for a while, but the loss from the system has been too great; life has been too largely dissipated, and what little remains can only flicker, is insufficient to revivify what remains of the worn-out fabric.

There is great thirst, of course, and it is useless to give the water called for, since it will be rejected as soon as down; but when the pill has had its effect, a spoonful at a time may be given, gradually increasing the quantity.

94. I have never seen a case that occurred after sunrise. I do not remember one that was not pre-

ceded by perfect health, a hearty meal before bed-time, sound sleep, out of which the patient is awak-ened by the attack between 11 P.M. and 2 or 3 A.M., rarely before or after these hours.

The foregoing description of what is now called *common cholera, or cholera morbus*, does not perfectly correspond with the disease so named in medical works. While I would not dare deny the correct-ness of the numerous authors who have classified and described the so-named disorder, I must say, for *myself*, that *I* have encountered none other than the form now imperfectly described.

To illustrate the foregoing, and make myself more accurately understood, I shall give a few cases.

95. 1st. That of Messire Beddard, a distinguished prelate of the Sulpician Seminary, in Montreal. He was taken ill, after his first sleep, in the night. The medical attendant of the house was called to his assistance, and prescribed as he saw fit. The violence of the attack was over in a few hours, although liquid dejections continued in small quantity for some time after. Hope, scarcely any doubt, was entertained of his recovery, after the storm had passed. But, although the discharges had ceased, he did not pro-gress in regaining strength. He rallied for a while, and again seemed less well. I was sent for about 40 hours after the attack. I found him dressed in his

4*

gown, standing in the middle of the room, surrounded by friends. He looked ghastly; with a husky voice he requested acid drink, craved for clotted milk, *du lait caillé*, above all other things. This his physician interdicted, and on this account I was called in. I approved of his taste, which I took to be an indication of what his stomach was capable of appropriating—a very grateful article of food, very digestible from the natural admixture of lactic acid with the curd, much consumed by the French Canadians. He ate some, with a little maple sugar and a few mouthfuls of bread; he wanted more, but was then restrained; it agreed well with the stomach; but he grew more and more feeble, became slightly delirious, and died some 50 hours after the attack, July 2d, 1825, aged 58 years.

96. 2d. Cammeron, Dorchester street, a healthy and strong man, ate a moderate supper of stewed mutton and potatoes (*left from dinner*) at 8 P.M.; went to bed at 10, feeling quite well. He was awakened at the end of his first sleep by an uneasy but painless sensation. Vomiting came, first of the stew he had eaten, *unaltered*, next of a mawkish, watery fluid, repeated in gushes as fast as it accumulated; the parallel of this took place by the bowels, urged by violent abdominal contractions, but no colicky pain or bile could be detected. Pulse scarcely perceptible

when I saw him at 2 A.M.; skin cold and clammy, some cramps in one leg, breathing easy and natural, with an occasional sighing respiration; voice much enfeebled; very restless. The first thing done was to deprive him of water, which he craved much, and rejected in gushes as soon as swallowed; the exertion of vomiting was a source of more exhaustion. From experience I well knew that a carminitive draught, from its mere bulk, could not be retained. One grain of solid opium was got down, and by much persuasion and close watching he was kept from tossing about, since this alone would keep up the irritability of the stomach and the retching. In half an hour's time the retchings had diminished, the pill remained down, he grew easy. A little later he became dozy, and in an hour was allowed to take a spoonful of water at intervals, gradually increasing the quantity, as it was thought the stomach could bear. All the trouble subsided into a good sleep, out of which he awoke nearly well, with the exception of thirst, a few muscular pains, and weakness. In twenty-four hours he had quite recovered.

Cammeron's case may be taken as a type of sporadic cholera that usually occurred in my practice in the hot summer months of Canada at Montreal; and the treatment I adopted was adopted by the very able and experienced Dr. Arnoldi, sen., and a few others.

I shall now give a third case, one in which no medical treatment was had recourse to until the malady had run its course and all discharges had ceased.

97. 3d. Barthélemy l'Espagnole, a native of Spain; a tall, lean, wiry man, about 60 years of age; he lived at the top of Sanguinette street, his house in a small garden, which he cultivated, and which supplied the wants of his wife and an only daughter; went to bed in perfect health on a Sunday night; was awakened with vomiting and purging, some cramps. He had no one else than his young daughter to send for assistance, through a long and lonely street, which he and his wife objected to. I saw him about 9 A.M., found him pulseless, cold, clammy, breathing easily, with an occasional deep drawn sigh; very thirsty. He had his senses, but was silent. This old man had very little substance to lose, was soon run out, and died.

98. 4th. One more case, since to me it appeared very interesting. Mrs. Cowing had just recovered from typhus of nearly five weeks' duration. Her convalescence was rapid, and, as is usual at such time, her appetite was extraordinary. On Sunday there was roast pork for dinner, of which she ate very heartily, and at 10 P.M., while in bed, she craved more, and was served with more. Her husband, a florid, robust and healthy man, ate also of the same pork at

dinner, at 2 P.M. In the evening he took his tea, etc., as usual, feeling quite well; retired to bed at 10 P.M.; about 2 A.M. he was awakened with vomiting, purging and cramps. The pork was the first matter rejected, *unchanged*, although it had remained in his stomach about twelve hours; all the other ingesta had *digested*. When I saw him he was nearly pulseless, cold and clammy; his florid cheeks were now slightly blue. Opium answered in his case like in those already described. His wife, who was only recovering from a protracted fever, did not experience the slightest disturbance from the same food that had acted so deleteriously on her husband.

99. This case is interesting, as showing how food will prove nearly poisonous at times to one person and not to another, particularly as regards the woman, who was still in a state of great debility, but whose appetite was voracious; whatever is craved for will suit the stomach, in preference to that which is not. This is another example of what hypothesists seized hold of in arguing against cholera contagion. They said: " If contagious, how come it that I, who was exposed, did not take the disorder ?" And in like manner Mrs. Cowing might have said, " Why did not the pork serve me in the same way, if pork was the cause of my husband's trouble ?"

Until 1817 cholera was unnoticed as being com-

municable. So was it in the sporadic cholera of
Canada; but when the cases had become numerous,
under what circumstances is unknown, it put on an
additional character, which it has maintained ever
since; and so it has ever been with every known contagion.

100. Up to 1832 I had carefully read what was
published of Indian cholera, and the fierce polemics
on contagion and non-contagion. The *printed* de-
scriptions of the cholera of India appeared to me to
be exactly like the sporadic cholera of Canada, and I
joined in the opinions of the non-contagionists. But
how soon was I undeceived! When I came to see
the imported cases, while the symptoms appeared to
differ only in degree, it was striking to the sight of
the beholder of a case that there was a great difference, but what difference *words* can not point out,
more than a witness can define what difference there
is between two countenances—each face possesses the
same parts, eyes, nose, mouth, etc., etc., yet a witness
can honestly swear to one person as positively differ-
ing from another; and so can any one say from the
first sight of a case of Asiatic cholera that it is not a
case of *common* cholera.

101. I have ventured to say, Sec. 92, that Asiatic
cholera is *not* a disease. I have often so stated to
medical men, to their astonishment and contradiction.

After listening to a long, lame physiological and even metaphysical argument, I have asked one simple question: *What is disease?* Until now I have not met with a single one of the ready argumentators who has been capable of answering the question. All writers and all remarks of medical men mention cholera as a *disease*, which expression I take to be an erroneous one when applied to this pestilence, and that it is a term which has warped the medical mind, by a mere word, into a vast variety of contradictory opinions, endless discussions, and absurd medication. It is accepted by the profession as a short and convenient way of designating the existence of something wrong, somewhere, in a part, or even in the whole system; but it conveys no idea of what this something is. On this account I have ventured on the difficult and new task, that of giving a definition to a word heretofore bandied about loosely, meaning something, but what that thing is no one has stopped to inquire. My effort in this respect is a new one, no doubt very imperfect, but it is hoped will excite the abilities of more able physiologists than the writer has any pretension to be, and that hereafter the phrase " diseased action" will be less flippantly uttered.

102. *Disease* is a disturbance in and among the *molecules* constituent of any tissue of the living body.

The natural function of the molecular matter is

perturbated, but is not destroyed. Disease does not arrest their *whole* action, only modifies it for a while, until the disease becomes exhausted, and is followed by recovery, or is arrested by death. During a diseased state the molecules still functionate more or less perfectly, but functionation is not suspended. Innervation, nutrition, secretion, and absorption occur; that is, new matter is formed, old structures are removed or greatly altered; deposits of fibrin or scrum or pus, or new organizations, take place; this is disease. None of these characteristics of disease are met with in cholera. Examples of diseased action: 1. The sensorium is obscured or over-excited, but perception continues, is not suppressed. 2. The lungs may be variously perturbated, but still continue to functionate—take in oxygen and give out carbonic acid gas and vapor. 3. The stomach may not chymify completely, but still does so to some extent. 4. The liver may be exceedingly disturbed, but the portal circulation is still carried on, and bile, such as it is, vitiated or otherwise, is secreted. 5. Kidneys may, like the liver, be much perturbated, greatly or slightly, but still eliminate more or less excrement from the body. 6. In inflammation the extreme vessels are perturbated, swell, even the structure containing them becomes infiltrated, but physiological action is not *completely* suspended. So on we might go through the whole catalogue of nosology.

103. How different in cholera! It alters suddenly all the physiological actions of the molecules it attacks, annihilates completely all function while its influence lasts. To illustrate the idea I entertain of the operation of the choleraic poison producing death without the intervention of " diseased action," I shall cite a few instances of death occurring in the complete absence of any possible disease. 1. A man falls from a height, is taken up apparently dead, and remains dead. In some of these cases, no bones are broken, no vessels ruptured, no apparent lesion can be detected; but he is dead: surely disease did not kill him. 2. One is shot through the heart and is instantly killed, before blood has had time to escape in any quantity to be accused as the cause of death; no disease acted here. 3. A man descends into a fermenting vat; he dies instantly, without disease; but physiologists of the schools say something about the arrest of oxygen reaching the lungs, etc., but cholera refutes the doctrine.   4. A drop of prussic acid put into the eye of a dog will kill instantly; here no disease has had time to be induced. So on we may go, citing numerous examples.

104. If we contemplate with close attention various cases of cholera, we shall find numerous examples of death taking place so suddenly that no possible disease had time to assist in the catastrophe. Some persons in perfect health have been stricken dead, it is reported, in fifteen minutes, even without discharges; but in all these sudden cases cyanosis was present. My case, cited Sec. 175, must have died very suddenly. If, on the other hand, we contemplate a less astounding case, we shall see that great numbers were severely attacked, and in twelve hours the colliquation had completed its work and left its victim *well;* nothing but weakness followed. Does disease act thus? I cannot afford to extend this small book further, else I might cite a large volume in support of my assertion, Cholera is not a disease. Reader! do not believe me, but learn for yourself, as I have done. The delusions created by education are delightful; so that, few things are more painful to the deluded than an attack upon, or a refutation of, their fond and long cherished errors.

105. It may be asked by some one: what good can a verbal distinction make? 1. In reply, I say the difference is not in words but in facts, as has been clearly shown; a wrong name leads to a misconception of the thing named. 2. The practitioner who calls cholera a disease will carry into practice his habits of treating disease where none exists, and work up for his guidance the idea of an imaginary pathology and a physiology which cholera utterly refutes. Out of this error in a name the wildest notions of medication have been adopted, useless in all cases, injurious in nearly all, and horribly cruel to the patient in many, as shall fully appear in the chapter on *Treatment*.

CHAPTER IV.

SYMPTOMS EXAMINED IN DETAIL.

Invasion of a patient. In all cases of cholera the symptoms, or rather the phenomena, are exactly the same ; they differ only in degree. In some slight attacks they do not all appear to take place, or rather are less noticeable; and in some rare cases the victim is at once prostrated and killed in a few minutes—he is overcome, and cannot seek or call for assistance, and dies where he happens to be. Such cases are reported, by respectable authority, to have ended in death in *fifteen* minutes.

106. The invasion is very variable as regards admonition or not; some have no previous warning, but the majority have. A gentleman, in the habit of taking a ride before breakfast, mounted his horse as usual, feeling in perfect health. He had not gone far when an uneasy feeling came over him; he sickened, and was conveyed back to his house; vomited copiously many times, purged the same without the slightest colic pains; his skin became clammy, wet, cold and cyanosed; had a few cramps in the legs; had taken brandy and opium; and by evening, a period of twelve hours, the colliquation had completely ceased; he slept well, and next morning had quite recovered, with the exception of being weak and thirsty.

107. The governor of Madras, Sir Thomas Munro, sat down to breakfast in perfect health, but was suddenly interrupted in his repast, being overcome in an indescribable way; he grew cold, skin clammy; said to his aide-de-camp, " I am an infected man, and shall die." He did die in ten hours. Men on sentinel duty, men at guard-mounting, who rose in the morning in perfect health, ate breakfast as usual, and marched with their accustomed smartness to the parade, were, without the least premonition, attacked and prostrated; men in their usual health were suddenly stricken while marching; mounted men, unconscious of any thing wrong in them, were suddenly dismounted. These cases were numerous, public, and officially reported. The same came under my own observation, as the case mentioned, Sec. 175 ; that of the soldier on a visit to his fresh arrived friends on the night of the 9th to 10th June; that of the old man's wife at Contrecour, and one or two in the Chesser family.

108. But the majority of cases had suffered a period of incubation after exposure, during which incubation, seldom reaching to four days, they remained perfectly

well, when at last the attack would occur as suddenly
as in all other cases. This interval of time suggested
the idea that the poison had to grow to a sufficient
amount, or sufficient virulence, before it could overcome the natural healthy action carried on in the
patient; when it arrived at that point the attack
became manifest as in every other case, sudden, more
or less violent, perhaps in conformity with the tem-
perament of the victim.

109. A regular attack takes place with an invariable set of symptoms of more or less severity,
preceded, in some cases, by an uneasy feeling of not
being quite right. The patient is soon prostrated; a
cold perspiration, or rather transudation, breaks out
over the whole body and limbs, which quickly assume
a leaden color, may deepen to blue, or even to brown,
according to the natural tint of the patient. The
fingers, toes, arms and legs quickly diminish in size;
the skin is wrinkled, shriveled, especially that of the
fingers, which are reduced in thickness one-third of
their previous size; the superficial veins are easily
seen as dark, or nearly black, lines; the temperature
is rapidly reduced *below* that of a dead subject; the
patient all the time complains of feeling hot. In all
this "cold stage," as it has been called by a supposed
analogy to fevers, there is never the slightest shivering, chattering of the teeth or trembling that is met

with in the cold stage of ague; on the contrary, the patient invariably complains of *heat*, wants exposure, and will not suffer covering, excepting of the lightest kind. The pulse soon becomes imperceptible at the wrist, and in a short time it cannot be felt as high up as the axilla, but what pulse can be felt is scarcely quickened in number; on listening, the heart is still heard acting feebly. Vomiting soon succeeds, and purging, by gushes, in astonishing quantity, of a *cold*, grayish-white fluid, resembling thin gruel, or "rice water," devoid of odor or taste (to the patient), and without colic; cramps, in the legs mostly, less frequently and severe in the arms, occur. The abdominal muscles also are cramped; but the cramp pain is less severe here than in the legs; the "bellies" of the muscles appear prominent between the tendinous intersections; the whole belly is drawn in. Inspiration is disturbed only slightly, in some few cases a little quickened, in all is voluntary; deep inspiration can largely expand the chest, and a long drawn sigh every now and then takes place, to end in a peculiar hollow moan. The expired air is quite *cold*, so is the mouth and tongue. The voice is peculiar and sepulchral in tone, approaching to aphonia. The eyes are deeply set in their orbits, and quite dry; no tears are produced even during the deep anxiety and anguish of friends and relatives present. The countenance becomes more cadaverous and ghastly than by death under any other form of disease.

Great restlessness prevails while the patient has strength to toss about on his bed; he lies mostly on his sides, curled up, until near death, when he may settle on his back. His mind is clear to the very last; he knows all that is passing, but is taciturn and speaks only in answer to questions. His perceptions are all natural, excepting that of external sensation, which he complains of as being hot, while he is cadaverously cold, and cannot bear the weight of covering. He complains of intense thirst, and incessantly calls for water, cold water, which is no sooner swallowed than is rejected with a gush. Urine is totally suppressed; but there is a constant desire, in the male, to void it. At last the patient becomes seemingly insensible—he is dying; but never is there heard a throat rattle; he dies quietly, without a pang or struggle.

A severe case, like that which is just described, generally dies in twelve, sometimes in eight, six, or even less hours.

110. Some such cases seem to have run and completed their course in eight to twelve hours; the infection or cataclysm appears to have exhausted itself, leaving the patient prostrated, with no apparent remains of the complaint, excepting the persistence of the dark color of the surface, but which gradually diminishes with the return of the circulation and the resumption of the pulmonary function of aeration, both of which had been completely arrested during the cyanosed state. Such a patient loses his restlessness and resumes his voice to some extent, gives hopes to his friends; but too often it happens that life only struggles on for a day or two more, when he sinks forever. Some of these cases resist death a while longer, and put on that peculiar state of the system said by many writers to be the "typhoid stage;" but this, on careful observation, presents nothing to justify such a name. This apparent fever is simply an effort to rid the system of a portion of the poison that remained after the undescribable evacuations had ceased and not carried all off, and is also an effort of nature at recovery.

111. A less severe form, but still of the collapse kind, is often met with, excepting of less severity and duration. Many of these cases are soon over; the discharges by the skin, stomach and bowels being very free; cyanosis is less deep, cramps less severe, emaciation less, breathing about the same, and aphonia; the expired air, tongue and surface are cold; pulse extinct at the wrist; urine suppressed, etc. Such a patient may be seen up and about the next day; he is only weak, easily tired, and much re-

5

duced in weight and plumpness; he is well. In this case it would appear that the poison, whatever it may be, was quickly all carried off by the discharges, or it had quite expended its force.

112. *In the mild cases*, cold and wet skin and cold tongue constitute part of the attack. The circulation is not completely arrested, hence cyanosis is absent, or nearly so; there may be some vomiting and purging, but no cramps, and the voice is scarcely altered; urine is not totally suppressed, but passed in small quantity only. Such cases may last longer than the second form. The prostration and loss of substance is slight. It would appear that in this last form the poison had made only a slight impression on the system, and is eliminated slowly.

113. *Cholarine*, a name given to a diarrhœa that is often met with some days *after* the great onset and force of an invasion has, to a considerable extent, exhausted itself. This name was invented by Parisian practitioners to distinguish such cases from real cholera. It is rarely met with in the first days or week of an invasion. The stools are feculent in appearance and odor, are liquid, brown, not frequent, and in small quantity, unaccompanied with colic, or only so in a slight degree. I have often considered these cases and had good reason to believe that many of them were due to mental anxiety acting on the bowels. This diarrhœa is easily controlled by a judicious use of opium, reasonable diet, and an assurance given to the patient that he is not affected with cholera; in this way the patients nearly all recover. But a certain number of real attacks are preceded by this derangement of the bowels, which, I feel confident, is not a part or necessary forerunner of an attack. Why should not cholera supervene on such a diarrhœa, when we see cholera attack persons laboring under acute diseases, as fever, typhus fever, even small-pox ?

114. Between the three degrees of cholera mentioned there are many shades of intensity, from that of an astounding attack that kills, without previous warning, and without discharges, in *fifteen* minutes or even less time; cases of great and sudden collapse; and cases in which there is only a simple uneasy feeling with moderate discharges. Hundreds have died before the cholera cause, the immediate cause, has had time to colliquesce the tissues into the particular and peculiar fluid which, being oozed into the stomach and bowels, excites the vomiting and purging usually met with, and also its transmission through the skin. In these astounding cases death has taken place in a few minutes' time in persons who have previously made no complaint of feeling ill. Marvelous as this statement may seem to the inexperienced, it is cor-

roborated by nearly all the writers on cholera in India, who, being military officers, and their patients of the same class, their opportunities for exact observation are greater than those of civil practitioners, who have to rely on the vague statements furnished by people in a state of confused excitement.

115. At the risk of being accused of repetition, the ' writer may be permitted to refer once more to what has already been said: " At Bundlecund, in the army of the M. of Hastings, the invasion was so sudden and violent that mounted men were stricken from their steeds, fell and died on the road; . . . . sentinels, after guard mounting and placed on their posts, were suddenly overcome, and quickly died." Such is only one of the official reports; the same has occurred again and again on numerous occasions in India, at Malta, and even in Montreal, as shall be referred to hereafter.

Let us now proceed to contemplate or examine the symptoms, or rather phenomena, that occur in this dire pestilence. It is difficult to decide on which function we should first inquire into, since they all depend on each other in the warm blooded vertebrate, *i. e.* the nervous, the sanguineous, and the aerative or pulmonary systems; besides, also, they often seem to be all attacked simultaneously.

116. What is understood by the word symptom, as met with in nosological authors, does not occur in cholera, excepting after the disorder has run its course, without death, and an effort is made by nature, as it is said, to restore the system to health; then comes that *quasi* fever, which has been supposed to resemble typhus, and is erroneously called by that name.

117. *The first symptom, or nervous one.* A general uneasiness is felt—"I am unwell." But this even is not always mentioned by the patient, for some died without making this premonitory remark. When the invasion of the malady is well under way, innervation is perverted in one respect only—the patient complains of heat, while he is as cold as a corpse; he objects to be covered, wants the bed clothing removed, the weight of which annoys him; a great restlessness prevails; he tosses or moves about on his bed; during all this time, and to the last, his intellect is undisturbed, and remains clear to the last moment that it can be manifested.

118. *The pulmonary system, or aeration.* Inspiration and expiration take place, to all appearance, as usual, excepting that there is occasionally a deep inspiration, followed by a similar expiration, resembling a sigh, and ending with something like a slight groan. The expired air, on striking the hand or face of the practitioner while examining the patient, feels like a cold blast—as cold as the atmosphere. It comes out of the lungs unaltered in temperature; and during the collapsed or cyanosed state, even when this lasts for several hours, its chemical constituents remain unaltered, or nearly so; the lungs have taken nothing from it, have given nothing back; that is, oxygen has not been absorbed, nor carbonic acid returned in exchange.

119. *Circulation, and the blood.* From the earliest examination of a choleraic patient the pulse will not be found much altered in time; but it is smaller and softer than usual, which alteration becomes more marked as the case progresses; gradually, it ceases to be perceptible at the wrist, next in the course of the humeral artery, and lastly in the axilla. It now becomes evident that the heart is incapable of sending the diminished column of blood on which it acts to the extremities of the arteries. Although the heart contracts and dilates with regularity, it does so like the heart of a frog or turtle, or other low animal, from an inherent power proper to it, independently, to a great extent, of the blood that enters it. There is a mere flux and reflux of blood from and to the ventricles—a mere remora of blood in motion. The column of blood is not driven to the extremities of the arteries and returned by the veins; in a word, circulation has ceased, and still the patient lives for some hours. His intellect remains unclouded to the last, in spite of theories that insist on the necessity of a full supply of arterial blood to the brain to support its function. Here we find two great functions essential to life in a well person arrested, or at least too much crippled, to supply the wants of the system they were created to uphold; and the absence of which, in any other condition of the economy than during cholera, results in speedy death. But in cholera the system seems to be indifferent to the suppression of these two great functions, and to the arrest of physiological actions, the integrity of which has, until now, been deemed essential to a living warm blooded animal.

The red globules of the blood cease to be acted on in the lungs, arterialized and crimsoned, and carbon given off; hence the leaden color of the body, and the stasis of the blood in the capillary system of the tissues.

120. *Digestion* is completely annihilated; consequently neither chyme nor chyle are produced, and no supply enters the circulation; besides, as circulation is arrested, no secretions can take place; hence it is that no urine is produced; that ulcers, acute or chronic (like the varicose) at once dry up; that the most virulent and copious clap at once ceases (to be resumed when cholera has run its course and the patient has recovered).

121. *Calorification.* Red blood, that is, blood containing red globules, is found in all vertebrate animals; in very small quantity in fishes, and more and more of it as we ascend in the scale towards warm blooded animals. In the lowest and lower grades, red blood is of no use to the animal, and appears to be one of rudimentary development in creation anticipatory of a higher future structure and want, not needed in the lower animals.

Many facts go to prove that it is the red globules that receive in the lungs the material creative of caloric, which, by these globules, is distributed to all parts of the body. In cholera it is self-evident that the globules undergo no change in the lungs; but, if they did, the effect of the change would amount to little, since the circulation of the blood is arrested, and these globules cannot be carried to the extremities of the vessels. To this peculiar condition of the blood and the arrest of circulation is the reduced temperature of the body greatly due; but not altogether, for the quick lowering of the temperature of the body of a choleraic patient is apparent in less time than could take place from a mere reduction of the temperature of the blood, and the long time a deficient circulation takes to permit a whole body to cool down to the ambient temperature. A choleraic patient becomes as cold as a subject dead several hours, or even colder, and that shortly after the attack; he cools down while *alive* more rapidly than does a body which had died of even a lingering disease, or one that has been killed.

122. *Cold.* This condition of a choleraic patient is no doubt due, in some degree, to the dark state of the blood; in other words, to deficient aeration of the red globules, and the diminished force with which the blood, as such, reaches the extreme arteries. In cases of arrested development of the heart, like in *puer cœruleus*, the temperature of the individual is always less than in those where perfect aeration takes place; and such persons, when exposed, quickly lose the little heat they possess. Another cause for reduced temperature is met with in paralysis. But both these examples do not illustrate or explain the rapid loss of caloric in a choleraic attack. The diminished or complete arrest of the circulation, even that of black blood, does not explain why, in cholera, the body is so soon cooled as it is. If this were due to the diminished circulation of even black blood, the length of time the circuit takes to be accomplished, and the number of times this circuit must be repeated before *all* the blood completes the round, requires more time to reduce the temperature by this means

5*

than it takes cholera to accomplish the task. A choleraic patient becomes cold in much less time than it takes a dead body, from any other cause, to cool to the same degree. Besides that, a body after ordinary death never sinks below the temperature of the atmosphere and the ambient bodies; on the contrary, a choleraic patient, lying in an atmosphere of 88° to 90°, is soon cooled down to less than 70° or 68°. His body is not only colder than the temperature of the room he is in, but the *interior* of his body is equally cold. This extraordinary fact is proved by his expiration being as cold as the air he inhaled; but his stools are still colder, and they come from the interior of his bowels.

123. During the cholera invasion of 1832, in Canada, I mentioned this fact to several practitioners, who seemed not to have noticed it, and affected not to credit my remark. It so happened that I was called to visit Mr. Horace Dickenson, aged 45, on the 21st of August, 1832, a wealthy gentleman, living in a fine house, situated in a very healthy locality. He had been suddenly attacked, was already in the cyanosed state, and purging still continued when I saw him. His family physician, Dr. Robertson, sat on the side of his bed, and was in the act of passing napkins under him to catch the evacuations that came by gushes. One poured over his hands; he looked

round at me with an air of astonishment, and exclaimed, "You are right; his stools are quite cold!" We spoke of this occurrence afterward. He said: "I did not quite credit your statement when you made it, some days ago; but now I am convinced of the reality." He went on to remark: "How could such large, gushing quantities of cold stools come out of a warm body?" This is only one instance of how superficially many practitioners examine and reflect on their cases. The body was not only cold outside, but cold throughout the viscera. A very little reflection would have drawn attention to this unusual subject, from the fact of a cold tongue, a temperature that could not well be less than that of the head or mouth that contained it; also, had attention been paid to the cold breath of a patient, in contrast to a warm breath breathed out in the coldest days of a Canadian winter, while in cholera, even in the hot summer days, say 88° to 90°, it is much below that of the inhaled atmosphere.

124. Where did this expired air meet with its reduced temperature, if not in the lungs; and if there, how could the lungs be colder than the rest of the interior of the body? These are only two instances indicative of the interior coldness of a living body of a choleraic patient.

125. A warm enema is soon returned cold, as has been noticed by others besides myself.

126. But now I must adduce a third, and an irrefragable proof of interior coldness, however startling the statement may prove to the readers of this sketch. To those who are unacquainted with the deep and perfect religious faith of the French Canadians in Lower Canada, all Catholics, what follows may possibly appear extraordinary, and perhaps censurable; but to a reasonable and generous minded man *all* religions are respectable, and should never be trifled with. Well, baptism is there esteemed essential to salvation, and to an inheritance of the kingdom of heaven. Many pregnant women, attacked with cholera—and, as repeatedly said before, the intellect of the patient remains unclouded to the last —many of these women, conscious of the certainty of their death, *urgently begged* that a Cæsarean operation should be performed, in order to save their fruit and give to it the benefit of salvation by "infant baptism." Several practitioners yielded to these entreaties and operated. I was present at a few such operations. Although I never once operated, I took advantage of the inevitable opportunity to pass my hand into the living abdomen among the intestines. In every case that I did so, the viscera felt quite cold. To prevent any doubt from being entertained on this point, I will give the names of some of the operators. The first on my list is my respectable and able colleague at the time, Dr. Pierre de Beaubien, resident physician, then in charge of the cholera hospitals, one of the most assiduous and pains-taking practitioners, at his post before six o'clock in the morning, and who, by his thorough anatomical knowledge and numerous carefully conducted autopsies, was able to throw much *negative* light on the pathology of cholera. Dr. Beaubien is still a practitioner in Montreal, and principal physician to the Hotel Dieu Hospital there. He, I believe, operated three times. Dr. John Stephenson (the first person operated on for staphyloraphy, by M. Roux, in Paris) operated on two. Dr. Munro, at present physician, along with Dr. Beaubien, at the Hotel Dieu Hospital, operated about ten times, as well as I can ascertain. Dr. Robertson operated on one; Dr. Vallée operated on six, to my own knowledge, besides others. As the medical gentleman I have named were, and are, of the first respectability in Montreal, honorable and conscientious in their practice, which was public, and their doing necessarily made known to my office of Health Commissioner, I have felt it to be no breach of confidence, but a duty, to name them while writing on cholera.

127. On the 27th of Oct., 1865, I had a conversation with Dr. Beaubien on this subject. He mentioned

that the fœtus was dead in every case he operated on, unless it might be one in which Dr. Munro assisted him, who thought he did see some slight indication of a movement, but that he, Dr. Beaubien, had doubts of the fact. We can easily conceive the existence of a passive movement occurring to a fœtus in removing it from its folded up position in the uterus; and I may add, without offering the slightest offense to Dr. Munro, that he was at that time a young man, and very zealous, conditions that might lead to error, in the hope of success.

In justice to Dr. Beaubien I must state that he told me his Cæsarean operations took place in the shed hospital, on women who had just expired, or about to expire; but that he had not operated on a woman actually alive. In a letter I have from Dr. Munro, he mentions having operated in all thirteen times, not all of them cholera cases, women on whom he was in attendance for the purpose of operating, and waited to nearly the last moment of life before operating. As to interior temperature he has no recollection, never having given it a thought.

The practice of Dr. Vallée was principally confined to the suburbs, among the laboring classes. Finding that many of the Cæsarean operations were performed too late, he was easily prevailed upon to operate while the patient *still lived*. It was in some of his cases I attended, at his and Madam Tavarnier's request, that I had the opportunity of passing my hand into the live abdomen, noticing the interior coldness of the viscera and that of the fœtus. In these cases the funis was simply divided, and no search made for the placenta, and no blood was discharged.

128. On more than one occasion I made use of a "physician's thermometer;" inserted it into the rectum. I also got Madam Tavarnier, a celebrated midwife in Montreal, to pass it into the vagina of some of the patients, especially into that of some of the females operated on by Dr. Vallée, at all of whose operations she assisted. She carefully noted the temperature, and made her report to me. At present I cannot give a copy of her report, which was lost, along with many of my notes, during the political troubles of 1837, etc.; but I remember well that the temperature indicated by the thermometer *in vagina* was lower than that of the atmosphere at the time, during some of the hot days of July and August.

These operations will be noticed again when treating of a few other curious, interesting and important facts connected with this remarkable and still obscure malady.

129. In cases of saline injections, which were at a temperature never less than 110°, at times 112°, even 120°, and of which injections a portion oozed into the

intestines and came away in gushes, these discharges were cold; therefore, the injection must have lost its high temperature in passing through the interior of the body.

I trust I have adduced sufficient proof that the interior of the body is cold, while the patient is still living, in cases of cholera.

130. *Vox cholerica*, or partial aphonia. After much reflection on this symptom I have been unable to arrive at any satisfactory explanation. To commence with the least satisfactory of my thoughts on the subject, a mere idea, was, how far the empty or contracted state of the innominata and the aorta could disturb the action of the recurrents, seeing that they wind round these vessels, which, in health, are permanently distended. I well remember the case of a would-be suicide who cut his throat, and the left recurrent, and who lived a few years after. His voice was weak and rough, and to produce it strong expiration was necessary; but still, in his case, the voice heard was not quite that noticed in cholera. The next, and more reasonable explanation, might be attributed to the altered state of the lungs, and the quality of the expired air. Another and more likely reason might be found in the deposition of some of the choleraic, pultaceous matter in the larynx and about the vocal cords; for it was not uncommon to find this matter deposited in small quantity in the trachea. These are mere ideas, only speculative, so that little, if any, reliance can be placed upon them.

131. *Perverted sensation of heat* on the surface and in the stomach is constantly present in every case, while, in every other respect, all other sensations, as hearing, sight, smell and taste are natural, with an undisturbed intellect. This cannot be explained by any known physiological theory, or even by the wildest speculations of would-be physiologists. We must remain satisfied with the fact, and not presume to explain that which is not given to us to know.

132. *Cramps.* Can the cold state of the patient, and the vacuity or uselessness of the arterial system, be related to their production? It is well known that many persons past thirty years of age will be seized with cramp in a foot or leg when lying down, and cold at the time. Another well known fact is, that after great and rapid loss of blood, or other quick drain from the body, cramps are not uncommon. No doubt there is some other influence in their production than these now mentioned. A few writers on Asiatic cholera state cramps as pathognomonic, even in slight cases. But there are many cases wherein this symptom has been absent, and they do not occur in children.

133. *Urine, suppression of.* It is easy to account

for this symptom. Vomiting, purging and transudation have completely drained the system, and whatever fluid may have been taken in is almost as soon
rejected, and none remains long enough in the stomach to be absorbed into the circulation to fill up the
deficiency, even if absorption could take place in the
cyanosed state of cholera; the arterial system is
empty and cannot be replenished for the reasons just
stated; besides, it is doubtful whether the impulse of
the heart can drive the little vitiated blood within its
power as far as to the extremities of the emulgent arteries; if this do not take place urine cannot be generated. It is generally said that secretion is derived
from the blood; but this is a loose assertion, repeated
by routinists, who seem not to be aware that the arteries are *common carriers* of blood and the fluids
poured into them by the process of ingestion; but it
is not from the blood, as such, that the materials of
secretions are derived; it is had from the heterogeneous materials added to it. This is not the place
to discuss these facts, which I have exposed many
years since. Since there is neither material to furnish secretion, nor circulation to carry the material
to the secretive organ, none can take place, and, of
course, the bladder remains empty.

134. It is stated that urea is found in the blood;
there is nothing remarkable in this, that a compound

element should be found in small quantity in the blood that remains, since it is unknown in what way the choleraic poison affects the chemical components of the tissues and fluids, and might spare this element, while by its mutation of ordinary compound elements into others of a different quality, only differently grouped, without losing or gaining a simple atom. One or two analyists have said that the quantity is great; no doubt, if it be spared at all, being found in blood greatly diminished in volume, the quantity would be disproportioned and appear large. Some persons may make this objection—since cholera permits the complete escape of the salts of the blood, why not urea also? The salts that do escape are mineral, while urea is a created body out of organic, not mineral, matter.

135. *Strangury,* as it is often called in these cases of ischuria, is almost a constant occurrence in male patients, a ceaseless but ineffectual desire to urinate; and I have known a busybody practitioner introduce a catheter to draw off what urine he supposed was in the bladder; of course, he found none. For nearly the three or four first weeks' continuation of cholera, I was unable to imagine what was the cause of this distressing symptom. At last I unexpectedly discovered it, a most simple one, which I shall now describe. It is easy to get a good view of the pelvis in a cholera subject, since the intestines are all empty, and take up but little space. They can be lifted or moved out of the pelvis; this affords a good view of the urinary bladder. But, first, there are those who suppose that in the operation of discharging the bladder it contracts in *all* directions down to almost complete vacuity. This is an error that lithotomists ought to know, even those practitioners who frequently draw off the urine with a catheter. Now, the base of the bladder lies against the floor of the pelvis, between the back of the pubis and the front of the rectum, in the male, where it is tied down to this floor, and is never removed thence, however much the viscus may be distended with urine; here it forms a *flat, adherent disc,* about two or two and a half inches in diameter, from side to side, and from before backwards between the pubis and rectum; in t he very centre of this disc the urethra opens. When t he bladder expands by accumulation of urine, it is the sides and summit that expand, and a portion of the base also stretches to some extent, but the base never leaves its attachment to the floor of the pelvis, to which it is affixed by pretty close cellular tissue. When the bladder is empty all contracts, the summit and sides as far as to the lateral limits of the base disc, and in this state the summit of the bladder forms another disc, of equal dimensions to the base one, and comes into immediate flat contact with the lower one. If, now, the intestines be pushed away, on looking into the pelvis, this summit will come into distinct view, with a depression in the centre, like an umbilicus, corresponding to the internal meatus that perforates the base disc. Let the anatomist, with a curved scissors, cut into the bladder all around the point of meeting of the two discs so as to separate them, and by taking the edge of the upper disc with a forceps gently pull it away from the lower one; he will find that the umbilicated depression is produced by a portion of the summit being forced *into the meatus*, acting in the latter like a foreign body, and creating the sensation of micturition. If this dissection be well conducted, on taking hold of one limb of the upper disc, and gradually drawing it away, it will be noticed that some little force is requisite to draw the invaginated portion out of the meatus. The complaint of incessant desire to urinate is most noticeable in those cases that have been accompanied with violent vomiting and abdominal contractions, along with gastric cramps, for it is these compressions that force the viscera against the summit of the bladder and press its centre into the meatus.

136. Females escape this accident, by reason of the different mode of attachment of the bladder, a considerable portion of which lies against the anterior surface of the uterus and anterior face of the vagina.

I trust I have not uselessly wearied the patience of the reader in this long detail; for I confess that, until I discovered this fact, I had not a very accurate notion of the extent of the attachment of the bladder, although I had often dissected these parts, and had many times performed lithotomy.

137. *Transpiration.* The discharge through the skin, could it all be collected, would amount to many pounds. It commences almost at the onset of the attack, and is always as cold as the surrounding atmosphere; in a hot summer day of 90° it is colder than the atmosphere and the furniture of the room. Most writers on cholera use the word " perspiration;" but perspiration, like secretion, has no existence during an attack of cholera. The discharge through the skin is not brought to this extensive organ by the extreme arteries for the impulse of the heart does not reach so far; all is stagnation.

To account for this great oozing through the skin, as well as that which is poured into the alimentary canal, both fluids being of the same quality, search must be directed elsewhere than to the vascular system, the operation of which is arrested in cholera. Writers on cholera say that these fluids consist of the serous parts of the blood, the red globules and part of the fibrin being left; but this does not explain the whole loss, nor how this serous fluid reached the surfaces to pass out, for neither the absorbents, the veins or arteries act; nor does the assertion (for it is merely an assertion) demonstrate that the " rice water" matter consists of serum, as such, and of fibrin, to which two bodies it has not the slightest resemblance; nor can organic chemistry prove any thing beyond the simple fact that it is animal matter, reducible by analysis to C. H. O. N. in special proportions, besides containing a few salts that are always present in the blood, and the tissues outside of the blood.

It is possessed of peculiar properties that are not found in any known secretions derived from the torrent of the circulation. It freely arrives at the cutaneous and intestinal surfaces and permeates them to escape from the system; but it never passes through the serous membranes, since none is ever found in the peritoneal, pleural, or arachnoid cavities. None is ever met with in the urinary or gall bladders, but some is occasionally found in the bronchi in small quantity. Whence, then, does this peculiar fluid come, and from what is it produced? are important questions.

138. *Catalysis* is the only theory which, to me, appears capable of rendering an explanation. The

cholera poison, whatever it may be, and however it
may enter the body, soon acts on the juices combined
with the tissues, and which, by a simple mutation of
the numerical proportions that constitute the compound elements which make up the different elements
of the body, creates another set of compound elements
which is fluid, and has the property of freely passing
through the tissues to reach the outward surfaces and
escape, dispensing with the services of the absorbent
and arterial systems.

139. *Bile.* The gall bladder is always distended
with natural looking bile, in every case, and the *duc-*
tus communis choledocus and the *ductus hapaticus* are
both perfectly open and unobstructed in their whole
length. How the bile remains in its bladder and is
not forced out into the duodenum and pumped into
the stomach by the violent vomiting of the patient,
and the strong contractions of the abdominal walls
compressing the cyst, baffles explanation; for, in
ordinary vomiting, bile is always, towards the last,
forced more or less into the stomach; while in cholera an invariable characteristic is that, during the
whole duration of the attack, no bile is ever discharged, none even after the acute stage has terminated and a return towards health has commenced.

However, the presence of bile in the gall bladder
after arterial action has mostly ceased, need cause no

surprise, since bile is the product of the splanchnic veins, and the liver, like many other portions of the system, continues to act after animal death. But that which remains unexplained is, how is it retained ?

140. *Leaden color, blue or cyanosed skin,* appears simple enough. A portion of the venous black blood, and, perhaps, a little furnished by the non-aerated arterial blood is not returned to the great vessels and heart, but lies stagnant in the *rete malpighi* of the skin, also among the vessels, and gives a very dark color to the abdominal viscera, as is seen in Cæsarean section, and which, seen in the abdomen, some writers have called *congestion*, a convenient word; but whether the blood, so stagnant, consists of globules escaped through the parieties of the vessels as whole globules, or is hæmatin, separated from broken up globules, is not yet clearly known.

141. In rapid cases, such as recover in twenty-four hours or less, the blue color almost as soon disappears, or nearly so; but, in cases that linger longer, the skin remains dark for several days, the blue color gradually changing to a dirty red, less and less so daily. However, some of the dusky red tint will be perceptible as late as the *tenth* day, as I saw in the case of a very fair skinned lady, who then died,

6

exhausted, in that state which has been erroneously called typhoid.

142. *Mental faculties.* During the whole of this wonderful commotion in the economy it is exceedingly interesting to notice that the mental faculties suffer only *one* aberration—an erroneous sensation of heat, while all is preternaturally cold; in all other respects the judgment remains perfectly sound and unclouded, so much so that the remark has been made within my hearing in regard to a weak minded or trivial person while healthy, that such an one has now sober senses for the first time.

Routine physiologists insist that the manifestations of mind from the brain require a full supply of healthy arterial blood; but in cholera the blood sent to the brain is disordered blood, and the quantity passing through it, like all over else, is much diminished. Verily, this disorder contests many opinions that have been handed down as true and promulgated as essential to animal existence.

143. *Thirst.* This is not due to fever, to inflammation, or to an excessive amount of salt in the blood, as a meal of salted food will excite the salt blood, calling for dilution; not due to fever, for there never is any; nor is it like the thirst that follows rapid loss of blood, and which is a voice calling for repletion. The cholera thirst quickly follows, almost precedes, the first evacuations, lasts until the attack has ceased, even during the "typhoid state," for several days, growing less and less urgent as the system becomes replenished. It is the most urgent, constant and distressing symptom; the craving for water, cold water, is incessant, and is no sooner drank than it is thrown up again, and more is called for; so irritable is the stomach, it does not remain long enough to be absorbed, even if absorption existed, which is not the case in the cyanosed state. This thirst is due to the great and rapid loss of fluid the system has suffered, and the craving is more the voice of nature than that of the patient, ever crying aloud, *I am empty, fill me;* and the cry will continue until the depleted system is restored to its usual state.

We have now come to the end of the critical examinations of the phenomena ever present in cholera—it is hoped not without exciting some interest in the reader.

CHAPTER V.

Departure of cholera from Great Britain, its arrival in Canada being the introduction of a new pest into a new country. At the close of Chapter I. it was said, "we shall trace it, led, as it were, by the hand, across the ocean into the New World."

144. By direction from the Colonial Office, London, the Governor, Lord Aylmer, sent a message to the House of Assembly, February 3d, 1832, recommending that a bill should be passed for quarantine and health purposes. About the 25th, the bill was reported and passed. It was the first sanitary and quarantine measure ever enacted in Canada. It empowered the Governor to name a Board of Health, to consist of senior magistrates, a Health Commissioner, a Resident Physician, establish a quarantine station at Grosse Isle, below Quebec. This act possessed two great merits: 1st, it was short; 2d, the Governor was to make the appointments, and could, of course, displace his appointees at pleasure.

145. Grosse Isle **was** an excellent station, **an island,** that could not be reached without permission, and had safe ports and good anchorage. Here he established a military post, with officers to command, and a small battery to enforce obedience to passing vessels; the Resident Physician on the island to attend to the sick; and the Health Commissioner was executive officer. Dr. Griffin, surgeon **of the 23d regi-**ment, was appointed to the latter office; being a military man, he was habituated to system, method and obedience, and to enforce the same. But his office was a new one, created in a hurry, and not well organized for some time. He took possession of his post on April 23d, only ten days before the merchant fleet came rushing in, this year later than usual. On the 1st May snow fell and covered the country "in the white robes of winter."

146. A similar Board of Health was created for the cities of Quebec and Montreal. The civil magistrates put on these boards were relied upon by the public as gentlemen of character and honor, and not place-holders. They simply gave countenance to the Executive of the Board, and never introduced impertinent interference in any district. Quebec was not quite as well off as Montreal, because the great importing merchants and consignees of vessels could exercise some unseen control, while Montreal escaped this, seeing that vessels which arrived here had passed through all necessary supervision.

147. Shipping from Great Britain endeavors to leave towards the end of March, so as to pass into the gulf and up the St. Lawrence before the descending ice accumulates to bar their passage. In 1831, the first arrivals at Quebec were as early as April 24th; but in 1832 this fleet was caught in the ice of the gulf April 20th, and some of the vessels choked therein for fifteen to eighteen days before working through; the first did not reach Grosse Isle before the 3d May. Let us stop here for a few moments to notice the departures of vessels from the ports of Great Britain and Ireland, infected places.

148. The Robert, from Cork, now at Grosse Isle, 14th May, a few days out, with passengers, had 10 deaths, the last one on the 23d April. Seven of the ten died of *common* cholera; three of Asiatic cholera.

Constantia, from Limerick, 28th April, 170 emigrants, lost 29 in fifteen days; 3 of Asiatic cholera, 11 *common* cholera, the rest diarrhœa.

Elizabeth, from Dublin, May 28th, 200 emigrants, lost 22; 2 of Asiatic cholera, 11 *common* cholera, and 5 of diarrhœa, a few days out.

Carrick, from Dublin, arrived at Grosse Isle June 3d; lost 42 in the first fifteen days out; all well on arrival.

Brig Brutus, from Cork, 270 emigrants, in a few days had many deaths; after that all well.

Ship Brutus, from Liverpool, 18th May, 330 passengers, mostly people from the agricultural districts. On the 28th, in the morning, all were well. In the afternoon had one man and one child attacked, and died. On Tuesday death made fearful ravages. On Saturday threw 13 overboard. On Monday the mate, steward and cook took ill. Captain made for Cork, but went into Liverpool. Total deaths since cholera broke out, 81. The foregoing are not selected instances of cholera breaking out on shipboard shortly after sailing for Canada, but were taken promiscuously from among many. It will at once strike the reader how similar are these facts to those already mentioned as attacking the British shipping in India, and the two French frigates after leaving Manilla. The reader will not fail to notice the glaring attempt at deception practiced by the captains on the quarantine authorities; such as, "the Elizabeth lost 22, only two of which were of Asiatic, and 11 were of *common* cholera, and 5 of diarrhœa" —nothing is said of the remaining 4. Where did the veracious captain acquire his great diagnostic knowledge between Asiatic and *common* cholera?

The Carrick, from Dublin, lost 42 in the first fifteen

days; after that all was well, until arrived at Grosse Isle, June 3d, when one woman was attacked.

The brig and the ship Brutus each speaks in the same way for itself.

The Elizabeth lost 22, of which 2 were of Asiatic and 11 of common cholera: what a knowing captain! But we have all heard of *"sea" invoices* being made by captains of vessels to suit emergencies.

149. It was said that "we shall trace cholera from England, the leap-off point in Europe, across the ocean into Canada." It is hoped that the shipping cases, just referred to, will suffice to establish that point. And that they will confirm what was said, Sec. 88, that on ship-board cholera would attack the crew and passengers in a short time, as early as twenty-four hours after leaving port, seldom later; and by the third or fourth day nearly all on board susceptible of the pestilence would be down with the malady; that by the end of ten days, or two weeks at most, scarcely a new case would happen. But when the ship would reach her destination, and her people get on shore, some who had escaped while on board might, in a new atmosphere, be attacked; and the *well* passenger would infect some of those they approached, as was the case at Mauritius.

Quarantine station of Grosse Isle received ships from British infected ports early in May, some of which

vessels had had cholera shortly after leaving the
infected port, and the pest was soon over on board.
These vessels reached the quarantine all well. They
were, however, detained for a few days, the passengers
sent on shore, they and the vessel well cleansed and
thoroughly aired, after which they were permitted to
go to the end of the voyage, Quebec.

150. It is curious to observe that the pest did not
appear in Quebec, the first stopping place, for the
space of an entire month, although several hundred
ships moored there, and many thousand passengers
had debarked. The first appearance of cholera took
place Friday, the 8th June, as was made public by the
Quebec Mercury of Saturday, the 9th. " Since yesterday morning eight cases have occurred and three
deaths, and two others despaired of. This disease first
appeared in a boarding house in Champlain street, a
narrow street between the base of the cape and the
port, kept by Roach. The patients are emigrants, and
said to be some of those landed on Thursday evening
from the steamer Voyageur. 3 P.M. Fifteen cases
have appeared since yesterday morning, seven of whom
have died."

151. This account is defective, in not stating where
the Voyageur picked up her passengers; but as there
was no intermediate place between Quebec and Grosse
Isle where any passengers could be had, we are forced

6*

to suppose that she got them at Grosse Isle, either openly or clandestinely. The dates of the arrival of the Carrick at quarantine on the 3d, and the outbreak in Champlain street on the 8th, are sufficiently precise to fix the importation of the pestilence on the Carrick.

All the newspapers that pander to the merchants for support at once railed out against the *Mercury* for creating " unnecessary " alarm. One paper said: " The *Mercury* is censurable for spreading such a report, creating an alarm that must affect *commerce* over the continent of America, and will prevent the summer influx of pleasure seekers from the U. S., who spend large sums of money in our cities." In consequence of the alarm so excited Dr. Morin, Health Commissioner at Quebec, and T. A. Young, Esq., Secretary of the Board of Health, went down to Grosse Isle to ascertain all the particulars respecting the crew and passengers of the Carrick. They made the following report:

152. "BOARD OF HEALTH, QUEBEC, JUNE 8, 1832. Various reports having been circulated that a vessel had arrived at Grosse Isle in which there were several persons ill of Asiatic cholera, PUBLIC NOTICE is hereby given that the Health Commissioner, having proceeded to Grosse Island, by order of the Board of Health, has reported that the brig Carrick, James

Hudson, master, from Dublin, arrived at quarantine on the 3d instant; that there were on board 133 passengers, all of whom had been landed, and are in the Emigrant Shed; that the vessel is undergoing the usual process of disinfection; and that at the time of his departure, on the evening of the 7th, there was not a case of Asiatic cholera on the island. By order of the Board of Health. T. A. YOUNG, Secretary."

153. This proclamation from the Board of Health is so studiously obscure that it excited doubts regarding the honesty of the signers, and increased the prevailing alarm instead of quieting it. It was in consequence of rumors of the existence of cholera that the Commissioner went to Grosse Isle, returned and made this ambiguous report, but not a word is said of what actually existed in the city! In a day or two more the pestilence had spread so widely that disguise could no longer be maintained, and, as a substitute, the cry was spread, " Cholera is not contagious, it is merely in the air, etc. Let not the people be alarmed; attend to business as usual; live frugally and temperately; observe great cleanliness. It is only the reckless, intemperate and drunkards that suffer." Abominable falsehoods like this were daily repeated for a while, and even announced from the pulpit, notwithstanding it was notorious that the affluent and most respectable members of society died daily.

Members of Parliament, magistrates, wealthy retired gentlemen, living in their own sumptuous houses, on their own property, in select places as to salubrity, were attacked and died as early as the poor and the profligate, and in a statistical proportion greater than that of the poor. Still the would-be moralizers and teachers of society did not perceive that they were all this time calumniating the character of the best men, even their friends, who filled the highest stations.

154. The first cases of cholera that broke out in Canada happened in an emigrant house kept by the man Roach, on the noon of June the 8th, among passengers brought there by the steamer Voyageur. This same steamer carried a load of emigrants from Quebec to Montreal, and arrived at the latter place late in the afternoon of the 9th. At once the passengers crowded upon the wharf with their trunks and baggage. They managed to get away in a short time to the entrance of the Lachine Canal, in the Ste. Anne suburbs. When the wharf became clear of the incumbrance, a man was seen lying on his back, dying. I happened to be passing at that moment, and took a look at him. I had him carted to the Roman Catholic dead-house. Here he was seen by four other medical men, one of whom, Dr. T. Arnoldi, opened a vein, and by much squeezing and kneading of the arm about a tablespoonful of very black

thick, cold blood was obtained. He soon died. We never could ascertain who he was.

155. Among the departed was an Irish family. They took lodgings in the upper room of a neat, clean and small hotel near the port. A soldier in the garrison was related to them; he got leave to spend the night with these his people. About 8 A.M., Sunday, the 10th June, I was sent for. I found a woman and a man of those arrived the preceding evening dead, and the soldier blue and dying. He was at once removed to the garrison, and was the first case of death from cholera that occurred there. This statement is remarkable in several points : 1st, the steamer that carried the first cases of cholera to Quebec was the same pest ship that brought these cases to Montreal and landed the dying man on the wharf; 2d, while there is nothing unusual for cholera to break out among persons who had gone through previous exposure, but who, on reaching a new atmosphere (Sec. 149) should be attacked, there is, 3d, the sudden and astounding invasion of a healthy, vigorous man, the soldier, that had never been exposed before, and who was attacked and died in about 12 hours. 4th. The time from the outbreak in Quebec to that in Montreal was barely thirty hours. 5th. The distance between Quebec and Montreal is about 200 miles, and over this long distance, thickly

inhabited on both shores of the St. Lawrence, cholera
made a single leap, without infecting a single village
or a single house between the two cities—the houses
all along the distance not over one to two acres
apart. Theorists said that cholera traveled in the
air—was not given out by the patient. In such a
case as this the air must have been very capricious.

156. On my return to breakfast after visiting the
patients in the hotel, a messenger was waiting for
me to go to a man in Sanguinette street, St. Lawrence
suburbs. There I found an old laboring man lying
at the point of death, his bed, as usual, among the
French Canadians, surrounded by women kneeling,
repeating the litany for the dying. He was blue,
cold, and soon died. He went to bed quite well; was
awakened about midnight, with vomiting and purging. On leaving this house I was met by a messenger, who requested me to go to St. Constant
street. Here I found another man, like the last one,
dying of cholera, and, as usual, surrounded by neighbors, praying and repeating the litany. Both these
men had spent the day previous at work on the beach,
and had joined with the curious to look at the dying
emigrant lying on the wharf. During this Sunday
many other cases broke out.

157. At this time, 1832, there scarcely existed
such a place as a " tenement house" in Montreal.

Each French Canadian had his own house, small, neat, clean and comfortable, with at least a yard, and often a small garden. They were all acquainted with each other, and mostly intermarried. From this close relationship a death could scarcely occur that the moribund was not surrounded by relatives and neighbors, praying. A happy people, living on little, but a sufficiency, and contented. This kind habit of attending to the dying exposed those who assisted to the contagion, and they carried it into their own families. Hence, if nothing else conspired, abundant means for the spread of the pest was established. It is, therefore, not surprising that by Tuesday morning, the 12th, there were accumulated in the Roman Catholic burying ground, St. Antoine suburbs, over 100 unburied; fame said 200. This excited great clamor—some one must be blamed—this fell on the Health Commissioner. I immediately prepared to burn all above ground; this made a still greater clamor; I was remonstrated with by busybodies on behalf of the suburbs, people who feared that the smoke would spread the pestilence. My reply was: dig graves, if you want the bodies buried—that is not my office; if graves are not furnished I shall burn. The gentlemen of the seminary addressed the people at the church door, and these turned out and dug trenches, 10 feet wide, 8 deep,

over 100 long.   The dead were closely packed there
in tiers three to four deep, and covered over with
earth, leaving the remainder of a trench to receive
new comers.   Several such trenches were filled from
first to last, and many respectable persons, incapable
of getting a private grave, were herein buried.   It
was different in the Protestant grave-yards—the
number of deaths being much less than among the
Catholics, as the latter made up the majority of the
population.

158. I must now return to the Voyageur, a pesti-
lent steamer, owned by speculators, whose morality
lay in profit.   As this vessel carried her live cargo
from Quebec to Montreal without stopping on the
way, no one between the two cities was infected.   It
was said that she threw overboard several dead on
the passage, but I was unable to establish the truth
of the accusation.   However, I discovered the following fact : after passing Sorel about a mile a feather
bed was thrown overboard, and floated down the
river; a man named Latour, a small farmer, and
occasionally a small meat butcher, saw the object.
He paddled out in his canoe, picked up the bed, took
it to his house at the point of the Island St. Ignace,
and began to dry it.   This man took cholera and
died in 12 hours; his wife also took the pest and
died.   An old man and his wife lived on their little

property a mile above the village of Contrecœur, where the river is several miles wide. He was out in his canoe fishing, when a raft came slowly floating along with the current. The captain of the raft spoke the old man and requested him to take one of his men, dead, ashore, and bury him on the beach. The old man had not heard of cholera, and took the body ashore and buried it; this was on Saturday afternoon. During the night the old man took ill and died; his wife also sickened, and on Sunday morning the passing neighbors, seeing the house shut, mentioned the fact to his nephew at the parish church. The nephew went to his uncle's house, found the old man dead, and the woman about to expire. After doing his duty to them he returned to his home, a farm in the "Second Range." He took ill and died, but the intermediate people escaped. A drover, French Canadian, left Sorel, then slightly infected (and never much), to go to the Eastern townships; on his way he had to pass through a dense, primitive and uninhabited forest, seven miles across, in the centre of which there was one of the little wayside taverns, the only house; here he halted about midnight, took some refreshment, and in an hour or two set out to complete his journey. The next day the innkeeper was attacked, and after him his wife; both died. If these are not cases of con-

tagion, what else can they be ? I will add one more
case, taken from an American paper: " Oneida
Castle, July 23d : The captain of a passing boat
hired an Indian to bury a man from on board.
The Indian was immediately after seized, and died;
five other Indians were also attacked; all of them
died."

159. All the above are cases of attack produced
by immediate intercourse with infected persons, at a
time when such a pestilence as Asiatic cholera was
unknown, and had never before existed in the
country. All sudden attacks, breaking out within
12 hours after exposure, and all of them astounding
and ending in death; all among people who lived,
each in his own house, comfortably and healthy.

160. A city newspaper, of June 16th, says: " Business seems paralyzed. Physicians and ministers in
vehicles ply with velocity day and night through all
parts of the city and suburbs; druggists and apothecaries keep their shops open all night. On Friday
morning the carts again appeared in the streets,
bearing two or more coffins each, some with lids
unfastened, and some corpses without coffins." This
was true. As regards coffins, I set carpenters to
work to make them out of rude boards. I had
several depots of these coarse boxes, and furnished
them on application, by a ticket, to all those who

called for one; among the number were many respectable persons who could pay for decent coffins,
but none could be had other than those I furnished;
they, therefore, called at the office for them. I
remember one poor Irish woman who had lost her
husband and could not call or send for a coffin; she put
her husband, doubled up, into a chest, hailed a passing
cart, and in this way sent the corpse to the public
trenches. It was no unusual thing for a carter of the
dead to call out, on passing a house, have you any
one to send to the burial ground? In this way he
would get from two to four at a load, and call on me
for his pay.

These remarks may appear irrelevant; but I give
them to show how dreadful the mortality was for the
first two weeks. There were a few wealthy persons
who procured a better style of coffin for themselves.

161. Leaving the appalling throng of cases occurring among the lower and a higher class of sufferers,
I shall now notice what occurred among the upper
class, who were all well off. On the 15th, only five
days after the outbreak, the Seignior Saveuse de
Beaujeu was attacked and died in about twelve hours.
He lived in his own house, a fine one, in a healthy
street, with his daughter and two old servants; the
rest of his family he sent to his seigniory, where they
escaped. Tuesday, June 26th, Miss Hervieux, a

young lady, called to keep her friend, Miss de Beaujeu, company. She was taken ill in the evening, about 10 o'clock, full dressed; but feeling much restraint from the presence of two or three young gentlemen, night-watchers, remained dressed, sitting in an arm-chair; she died about daylight, the 27th. At once a coffin was ordered, a passing cart called, and immediately sent off to the burial ground, her family not knowing what had occurred. I mention this case since it created much extravagant talk then, and which took some years to subside; also because it was affirmed that she was buried alive, in the dress she died in, and her jewelry on her. To confirm the statement of her being alive, it was stated that she moved, or twitched, on being put into the coffin. This case, if true, is the only one noticed in Montreal of a choleraic corpse moving after death, as was mentioned in European journals to have occurred.

162. On the 15th, in the night, died, at one of the cholera dispensaries, Grant Struthers, a medical student, and next day Dr. Smith, and Pierre Delorme, a rich man. Benjamin Thatcher, who left in the evening to go to Vermont to escape the pest, died on the road in the morning. On Monday, Campbell Sweeney, Esq., and Mrs. Try; the latter was left alone in her room with her servants. Her house was situated in the centre of a large orchard, at a distance from other houses, spacious, airy and luxurious. 20th, Harry McKenzie, Esq., N. W. Company, rich, inhabited his own fine house; Stephen Sewell, Esq., Solicitor-General. A young lawyer, Levi Adams, recently married, lived in a new house, all of the furniture of which was new, in a remarkably healthy locality, St. F. Xavier street, adjoining the seminary garden—he died; the next day his wife and the servant girl died —a whole family blotted out in two days. Horace Dickenson, wealthy, in an airy house, only a few hours ill. June 21st, Jean Bouthillier, Esq., Sen., and his daughter, Mrs. Panet; Pierre Beaudry, Esq., and his son, living in a large orchard, dry, healthy, and very wealthy, both died; my much esteemed and wealthy friend, Benjamin Beaubien, Esq., attorney; John Flemming, Esq., President of the Montreal Bank; Miss Moffatt, aged 15, daughter of Geo. Moffatt, Esq., the largest importing merchant in the city; Mary Coton, of Tamworth, England. On the 7th August, Thomas Gibb, merchant, and many others. 26th, Rev. Newglove, at the residence of Wm. Lunn, Esq.; residence wealthy, beautiful and salubrious. These few, out of a great number of cases I might cite, will suffice to show that cholera spared not the affluent more than the lower orders; and as regards intemperance, there could be no greater libel uttered than even to hint at such a vice existing among people of their rank and respectability. If a statistic of all the cases were taken, it would show that the proportion of deaths among the wealthy and well conducted people was proportionately greater than among the lower orders; and as to cleanliness, the habitations of the suburbs people, of course not opulent, were cleanly, comfortable, and well supplied with wholesome living, in a degree and quantity sufficient for the wants of a modest and contented people.

163. By the 12th we managed to get into some system and organization. Temporary hospitals (sheds) were built; a steward, cooks, nurses, etc., located therein, under the excellent medical attendance of the Resident Physician, Dr. de Beaubien, and several advanced students who resided in an adjoining building. These sheds were got up in a hurry, and were, of course, very imperfect in structure and furniture for the first few days, because it was no easy matter to get workmen to expose themselves in such a pestiferous place. For want of cots and beds, straw was thickly laid on the floor as a hurried means of some comfort. As usual in such times, an ignorant newspaper editor (a pest to society) and a pair of intermeddling, discontented parsons made *one* visit to the place—*a short one,* for they were fearful of their own persons—and early the next morning their paper came out with

venomous diatribes: "the sheds ought to be called
slaughter houses; we found the patients lying on
straw on the floor, no bedsteads, no beds; we turned
away in disgust!" How easy it is to censure, when
one is under the protection of the "liberty of the
press," and comfortably seated out of responsibility
and danger! Had the libelers made a true statement, they should have stated that the condition of
the shed was the best that the anxious authorities
could procure in so short a time—a time of panic.
As soon as workmen could be had, trestle bedsteads
were supplied, palliasses, sheets and pillows; but
these guardians of the public never returned to see
the "slaughter houses," and to withdraw their libel.
Wine, brandy, cordials, condiments, the best of fresh
meat and poultry, were supplied without stint, also
unremitting care. The Resident Physician was at
his post at five o'clock in the morning; the students
remained *night and day*, assiduously administering to
the wants and comforts of the patients. Did the edi-
tors and psalmodists "do likewise?" Only once, and
that was to make a "sensation paragraph." The only
clergymen unfearful and unremitting in attendance
were the Sulpician Roman Catholic clergy; occasionally, but very rarely, a Protestant Episcopal, and
one Presbyterian clergyman, paid a hurried visit to
the "slaughter house." Dr. de Beaubien at once es-

tablished a burial ground at a distance on the common, kept graves ready open to receive the dead, and buried them as soon after death as possible. He used chloride of lime freely there, in and about the hospital. Almost every one that died he carefully examined, or had the autopsies conducted by his assistant students, as soon as no doubt remained of death, and carefully observed what the scalpel could reveal. I feel it to be my duty to bear the above testimony to the meritorious conduct of the parties concerned, although I have had no intercourse with them since thirty years.

164. The next sanitary step to meet the anxious wants of the public, but of little real use, was to establish a number of dispensaries, under the care of trusty students who had instructions for their guidance. These places were supplied with necessary drugs, and given to all who applied, night and day, gratuitously. Whatever good they may have done, one thing is certain, it contented the people.

165. At the outbreak of cholera Montreal had a population of 32,000 souls and thirty licensed physicians and surgeons, two small schools of medicine and a number of students. On the regular practitioners devolved all the duties of medical attendance, about one practitioner to a thousand well and ill persons, if equally divided, which was not the case.

There were three hospitals—the long established Hotel Dieu, the Gray Nuns' Infirmary, and a new and excellent establishment, the Montreal General Hospital, well conducted. Such were the only medical means to meet an appalling pestilence, hitherto unknown to Canadian practitioners, and who, like all over elsewhere, supposed the plague to be a *disease*, and consequently brought to their aid the dogmas of the schools—physiological axioms that had no bearing on the complaint, and which cholera refuted. The *methodus medendi* founded on such principles had no application here, proved useless in every case, and injurious in many. But scholastic education enables the weak minded to argue with an apparent display of real knowledge, and to override and subdue common sense.

166. The town of Three Rivers, half way between Quebec and Montreal, forbid steamers to come into their port, by which restriction the inhabitants escaped for a while, until some retrograde traveler from Sorel, above, brought the pest among them; but it did not spread much. Sorel, a steamer port, was early affected, but had few cases. This is easily explained, as that borough contains a divided population that keeps each aloof from the other, and is very unsociable. On both banks of the St. Lawrence, with a dense population, scarcely a case occurred

7

before six weeks, because no direct communication with Quebec existed—and no travel in the agricultural season takes place, excepting that of going once a week to the parish church.

167. From Montreal it was different: the great influx of emigrants was forwarded away by the Emigrant Society as fast as they arrived, and by them the pestilence was sown at each stopping place. Lachine first, then The Cedars, next Coteau-du-Lac, Cornwall, Prescott, Kingston, Toronto, Niagara, etc., to Sandwich, in Upper Canada, and to the opposite towns in the State of New York, spreading from town to town until it reached New Orleans. From Montreal to St. John's, thence through Lake Champlain, infecting the New York towns of Champlain, Chazy, Plattsburgh and White Hall, to Albany; on the Vermont side attacking Burlington, and extending to the interior towns of the State; but in all these places the great fury of the invasion seemed to have already expended its greatest force in Lower Canada. Although I have the dates of the successive marches of the pestilence from Montreal through Upper Canada, and into the United States to the Atlantic sea-board, and down to New Orleans, I abstain giving the calendar, that I may not overload this small book.

168. However, I must notice the march of cholera

up the Ottawa, because the itinerary is exact and particular. In 1832 the banks of this river were still lined with the primeval forests, and small settlements existed only here and there at special landing places, as the current is frequently interrupted by rapids, cascades, and unavoidable " portages."

169. Travelers up the Ottawa carried cholera to Carillon, Greenville, Point Fortune, Fox Point, on one side of the river, on the other to Rigaud, New Longueil, Plantagenet, etc., to Bytown, the terminus of travel, which place it reached 5th July, brought there by a woman passenger on steamer Shannon; she died the day after landing. Next day died the landlord of the house she was taken to, also the people in attendance on her and him; 88 died. At Rigaud there died 80, from the 8th July to 9th August. On the 10th Dr. Teasdale died. New Edinburgh, near Bytown, was dreadfully ravaged. At Plantagenet, Ottawa, died Mr. M'Kay, a young man in the service of Mr. Chester, 14 miles away. His mother went to his relief. On her return she took ill, and died in a few hours. Mr. Chester, Sen., took ill at once and died; Mrs. Chester died. Many neighbors attended her funeral, took ill, and all of them died; Miss Chester died ; her brother, Guy Chester, died, and a sister, Mrs. Molloy, died.

170. When it is considered that this was a new,

a forest country, and remarkably healthy, and that the pest was brought there and immediately scourged the people who communicated with each other, and to none other, we leave to the theorist to find and explain any other cause than contagion for what is now mentioned.

171. Opposite Lachine is the Iroquois village of Caughnawaga. Here 88 Indians died. These people were much exposed to the pest being brought among them from the occupation of the men, which was that of piloting rafts through the rapids. At the Lake of Two Mountains is a village of Iroquois, and an adjoining village of Algonquins; but here only two died. The explanation is easy. As soon as summer is set in the most of the people depart to the hunting grounds, and the few, with women, that remain, do not travel about ; nor is their place a thoroughfare of travel, as is Chateauguay, where, from the 14th June to 16th of August, 146 died.

172. A case of *second* attack took place in Montreal August 13th, and ought not to be passed over in silence. Mr. Alexander Gray, of the firm of Gray & Co., auctioneers, aged 31, took ill on Sunday and died early on Monday, the 13th. He had quite recovered from a former attack, and attended to business more than a week, when he was suddenly killed by this second attack. There were other instances of

a second attack, but it is sufficient to notice one by
name and date.

173. August 9th. Rev. David Hughes, Unitarian
minister of Yeoville, Somersetshire, England, aged
47 : on Thursday he left his family, in perfect health,
in company with his eldest daughter, to go to Upper
Canada. While on the boat, from Lachine to the
Cascades, he was ill; there were no berths, and he
had to recline as best he could. At The Cedars they
took the stage to Coteau-du-Lac; by this time he suffered much from vomiting and purging, but could not
prevail on the driver to stop more than once or twice.
When he got to the Coteau a doctor said he must not
proceed. He continued cramped for some hours, was
deeply blue, and died at 8 P.M.; was conscious to the
last, and was hurriedly buried in the night. Now
comes the important part of the case. His distressed
daughter returned to Montreal, and either took back
or sent back his luggage. Among this was a carpet-
bag containing some of the foul clothes he had worn.
This carpet-bag was taken into the garret of the house
where he had resided in Montreal, Mr. Ariel Bowman's, bookseller, St. F. Xavier street, and hung up.
The Sunday following, his youngest daughter, who
remained behind at Mr. Bowman's, went to play with
two of Mr. Bowman's children. They took down the
bag, overhauled the contents; in a vest pocket Miss

Hughes found a copper cent, took it as a remembrance of her father. This was Sunday afternoon—the whole household perfectly healthy at the time. At 4 A.M.. 20th Aug., Monday, I was sent for to see two of Mr. Bowman's children; they were both choleraic—very bad; these were his two children who had overhauled the carpet-bag. After prescribing for these two, a message came for me to go to another part of the house; there I found Mr. Hughes' second daughter, Sarah Hughes, 11 or 12 years of age, very ill of cholera, of which she died in two or three hours more. This girl went to bed perfectly well, and could not have been ill over seven hours. Mr. Bowman's two daughters recovered. These three cases are interesting, as affording an instance of cholera derived from "infected goods."* Sept. 18th, Mrs. Bowman, wife of Ariel Bowman, took cholera and died. Heretofore no cholera existed in this salubrious street, excepting the cases of Levi Adams, his wife and maid, on the 20th June.

174. The garrison of Montreal consisted of a part of the 15th and a few of the 27th regiments, some of the men doing outpost duty not far off—450 men in all; of these, 46 died. On the 19th June the colonel encamped his men, under canvas, on the Island St, Helen, maintained strict exclusion and seclusion and perfect hygiene; the result was that no more deaths occurred among them.

I must notice a few exceptional cases. It will be seen that a large number of the attacks were, without the least premonition, of an astounding character— killed in a few hours; but a good many of these attacks seemed to have at once purged the system of the poison, and such patients recovered in twelve to eighteen hours. This was the case with Mr. Gray, the auctioneer. (See Sec. 172.)

175. On the night of 15th June, a man whom I knew by sight, but not by name nor residence, came into the office, about 12 o'clock; when it came to his turn I asked what he wanted; he said he came for a coffin for his wife. I gave him an order for one, with which he went off. Before 1 A.M., the same night, a close and very hot one, among others a man came on the same kind of errand, and said, casually, that he passed a man lying on the pavement opposite the English Episcopal Church, not over 2,000 yards from my office. At once I sent to his relief. It turned out that the man was dead, cold, and very blue, but there was no appearance of his having vomited; in his vest pocket was found the order given to him for a coffin for his wife, less than an hour before. The body was sent to the dead-house, Place d'Armes, and thence in the morning to the trenches. I did not know his name, and believe that his friends, if he had any, remained ignorant of his sudden death. This case is cited to show how sudden some of the deaths were.

176. *A horse.* John Armstrong, master tanner, lived in a two-story house on St. Antoine Bridge. One of his men took ill of cholera, and for convenience' sake lay on a buffalo-robe on the floor; he soon died. The robe was taken to the stable and thrown into a manger next to the horse, a very fine one. Next day I was called to see Armstrong, ill of cholera; some one came in and said the horse was choleraic. I went into the stable, found the horse standing up, drooping, his ears hanging, his skin damp and sticky to the hand, mouth and tongue cold, breath cold, eyes dull and undistended; incessant purging of choleraic matter, the ingesta of food having previously passed away; there was no vomiting, because the structure of the cardia in the horse does not admit of vomiting. The horse soon died.

177. Poultry died in some places in the St. Antoine and Quebec suburbs of the city, with all the appearance of cholera. Dogs were also reported to me as having died of cholera. But on neither of these instances do I set much value, for fowls are subject to a disease resembling cholera, and dogs, in hot summers, are very subject to vomiting, and diarrhœa, and convulsions. However, in India the reports

e animals died off in great numbers;
ourg, Prussia, it is asserted that fish in
s died of cholera (!)—from the single
erberg 40 tons were buried. But it is
at many poisons mixing in the water
r kill the fish it comes in contact with.
of these deleterious substances, and
ill empoison a large extent of water,
· fish it comes in contact with. This
le after the great fire at Mirimachi,
. miles of forest, about forty years ago;
ed the alkali into the bay, completely
the salmon of that year, and greatly
distress of the people. Swine, it is
lera in Ireland and died.
orrect register of the numbers of inter-
Roman Catholic burial ground was
, none could be made; for during nearly
he carters carried the dead to the
knowing or caring to know the name
x of the body. They picked up, as it
ies as they went along the streets, or
r to cart away; they were common
could not even read, and, of course,
le of making a list, beyond noting the
carted, to draw pay for so doing. At
t burial ground affairs were not so bad,

7*

as the numbers taken to them were much less, and
most of them of a better class, who had friends to
attend to the burials.

179. It is, perhaps, well to say a few words about
the Board of Health, consisting of ten to fifteen
magistrates and gentlemen—a body quite outside of
the Medical Commission, and having only advisory
powers, but no right to interfere; nor did they once
intermeddle or give us any trouble. They appointed
a secretary, at a small salary. This gentleman, J.
Guthrie Scott, Esq., attorney, was a very timid man,
and never once left his house to attend the meetings
of the board. His office appeared to be that of col-
lecting daily reports of cases and deaths; but he
never once sought for the reports himself. He
managed to get occasionally a partial statement by
sending a note for it, and send this imperfect docu-
ment to the newspapers. I find it necessary to note
this, lest any one who may look over the files of
papers for this period should be led into the very
contradictory errors recorded. Mr. Scott was, as
above said, a very timid person. He had recourse
too largely to brandy to keep out the pestilence; it
overcome him and his judgment, and, becoming a
habit, soon terminated his existence. With this
exception he was an unblemished gentleman.

As examples of the reliance to be placed on Mr. Scott's " Reports," take the following:

"June 13—Cases,   94.   Deaths,   23.

June   15—Cases, 1,204.   Deaths, 230.

Up to 16—Cases,   —   Deaths, 102.

" J. G. SCOTT, Sec."

The editor of *Canadian Courant*, Sept. 5th, says: total cases, 4,385; deaths, 1,853.   The editor goes on to say, on his own account, at least 2,000 have died; and again he says, from June 10th to September 18th, deaths, 3,151.   Where Mr. Scott got his numbers I cannot imagine; but I notice in the newspapers that he attached my name to some of his reports.   It is certain he never once called to see me nor sent to me for a report.   Nor can I say more in favor of the *Canadian Courant*.   Both these sources, and all the published accounts, are equally erroneous.   *Canadian Courant*, 20th, says, many shops in St. Paul street are shut up.   A great many persons have fled the city to escape, and were seized with cholera on their road to an imagined place of safety.

180. The Hotel Dieu Hospital, a cloister, had only one case; the Congregation Nunnery had none; the Gray Nunnery Infirmary had only two; and the jail only one (perhaps two).   A number of doctors and apothecaries died, and several priests in the parishes.   One paper estimated that the total number of deaths in Lower Canada, with a population of 500,000, exceeded that of Great Britain, with a population of 15,000,000.

181. For the first fifteen days cholera was principally confined to the two cities of Montreal and Quebec, while both shores of the St. Lawrence between these places nearly escaped. But, at last, the pest spread from Quebec to Point Levi, opposite, and was gradually extended from parish to parish; many of these suffered, statistically, greater mortality than did the two cities.

182. From my notes, and disbursements to carters and for coffins, I venture to say that the deaths in Montreal reached quite to 4,000, if they did not exceed that number a little. By September 29th cholera had ceased with us.

183. A few words only will suffice for the cholera of 1834. It did not appear with us until 11th July, and ceased by the end of August. While we had some cases of astounding cholera, the number was less than in 1832, and taken altogether it was a much less disastrous plague; but it extended like in 1832, from person to person and place to place, in no instance until a contaminated person had preceded the outbreak.

Of Quebec, with a population of 28,000, I shall speak only as I derived information from the newspapers, one of which was always considered reliable, Neilson's *Gazette*. June 27th, the *Gazette* states the deaths in Quebec to that date at 1,200. June 8th to July 3d, 1,421 deaths; to July 18th, 1,662. " No correct estimate of public health can be made from the reports. In the hospital there have been 943 cases, and 590 deaths. But this hospital received many emigrants; this was not the case in Montreal. The number of deaths are more likely to exceed than be under 4,000."

The Quebec papers teemed with censures of the cholera authorities—with what justice I cannot say, as my office was confined to Montreal. I have no doubt they often expressed no more than the personal feeling of the writers, and some excuse must be made for excitement during a public calamity, in which newspaper scribblers notoriously take great liberties with persons who are otherwise occupied, and who scorn to reply to attacks made in the safe quarters of a newspaper office.

CHAPTER VI.

PATHOLOGY.

184. Let us begin with the blood—not that it is the first thing disturbed, but because we must begin somewhere; not that any change in its appearance is a cause of cholera, for its change is due to the choleraic poison; not because it plays any rôle of importance, for all its action is absolutely negative—it does nothing; it ceases to perform even its allotted office of calorification. Its stagnation on the surface is what first attracts our attention; and if we look into the interior, before or after death, we find the same stagnation, excepting in the larger vessels, where it only fluctuates to and fro, but does not *circulate* in severe cases.

185. Here we at once come into collision with that rough and superficial physiology taught while studying descriptive anatomy. 1st. It is daily taught in schools, and published in systems, by professors who initiate the student into the recondite but as yet imperfect science of medicine, that blood

is essential to animal existence; this doctrine is most
untrue.   2d. It is taught, and believed by superficial
practitioners, that the tissues are elaborated out of
blood; this is untrue.   3d. Consequently, that there
is a daily consumption of much blood; this is untrue.
4th. It is taught that, to supply the consumption,
more blood must be made *daily;* this is untrue.   5th.
It is taught that food chymified and chylified is
the pabulum out of which blood is made; this is untrue.   6th. Impressed with this roundabout way of
making blood, which blood so made daily is to cre-
ate and be changed into tissue, the teacher has imagined that the chyle, in its progress towards mixing
with *real* blood, gradually becomes tinged red,
"*almost evidently*" so, while yet in the thoracic duct,
when it has reached near the subclavian vein; this is
also untrue.   My belief is, in this instance, like a
thousand others, the teacher, imbued with the idea of
a necessity of blood, has persuaded himself that he has
seen the progressive steps of that creation.   As early
as in 1815, and for years after, I undertook many
physiological researches on digestion and on this
supposed creation and use of blood; but I was dis-
appointed.   I fed many cats, dogs, pigs, etc., and examined the chylopoetic system at various stages of
digestion; but my disappointment was great in not
meeting with what I had been taught to believe.   I

rarely met with even a tinge of red in the thoracic
duct. This was one of my first sources of skepticism
of the truth of doctrines. It is only yesterday, as it
were, that a contributer to the Philosophical Transactions managed to get accepted a paper, and a drawing to illustrate it, that he observed, in the micro-
scope, a blood globule unroll itself into a fibre to become muscular tissue! I believe his name was New-
port, but I have not the Transactions to refer to.

The foregoing digression I think proper to make,
so that I may not be misconceived in what I have to
say about blood in cholera.

186. *Choleraic blood*, in the cyanosed state, or in a
protracted case, or shortly after death, is found to be
of so deep a purple color as to appear nearly black.
It is viscid from the absence of serum and much of
the fibrin. The interpretation is, that while the
diluting serous fluid has been drawn off, the globules
have remained, concentrated in number, and dark
from the absence of aeration. The anatomical character of the globule is not altered; it preserves the disk
form, when taken out of the sinuses and great vessels;
but, if the blood be examined in the capillaries, it will
be found that many globules are diffused *extra vaso-
rum;* and that, in a protracted case, many will be
found either broken down, or that their hæmatin has
escaped, so that the *rete mucosa malpighi* becomes

permanently stained, which is the cause of the protracted color seen in cases called the typhoid stage.

187. *The heart and great vessels* near to it are all
filled with thick viscid black blood, which blood contains an amount of globules vastly in proportional
excess of ordinary blood. The right ventricle and
auricle are always full, and the left is nearly in the
same state. The pulmonary artery and vein show no
difference in the quantity of the blood they contain.
All the arteries contain some black blood. This
should be anticipated by a physiologist well acquainted with the movement of the blood in these
vessels during life, also *after* death. It is stated by
professors that the arteries are empty after death;
true—excepting in special cases reported as sudden death, and death from lightning; but such cases
I have not examined, and do not know the fact from
my own knowledge. I shall now stop a moment
to say a few words on empty arteries after death.

188. In disease the course of the blood through
the vessels does not cease at death, at the arrest of
the impulse of the heart on the arterial column. The
veins continue to deplete the arteries of their blood
for several hours after what is called death has taken
place. This depletion is due to a suction property
in the extreme veins, a power so great that all the
blood in the arteries is drawn out, and the round cal-

iber of the arteries is forced into a flat cord, in spite of the strong resistance of the arterial coat, ceases by tending to resume the cylindrical form, but cannot do so unless something enter to replace the vacuum which the venous suction has created. If a large artery, like the carotid or the femoral, be carefully laid bare, it will be found empty, and of course *flat;* but, as soon as a cut is made into it, admitting a rush of air, the vessel will at once expand and become an empty cylinder. This I have taught many years ago, and should not now recur to it were it not that not a word on the cause of empty arteries is to be found in *class-books.*

189. In cholera the veins quickly cease to act, draw no longer on the arteries, and consequently do not influence the motion of what blood remains in the arteries, which is necessarily black, since the lungs cease to aerate the blood many hours before death; the heart still acting, causes only a flux and reflux of the blood nearest to it, but effects *no circulation.* Under these extraordinary circumstances, it must be expected that blood of a black color will be found in the arteries.

190. It is curious to see both arteries and veins equally filled with black blood, coursing languidly through the mesentery during life, as I have seen in

the Cæsarean sections performed with the delusive
hope of saving the life of a fœtus.

191. *Cavity of the skull and the brain* are the next
places where pathological change has often been asserted to exist. When the calvarium is removed,
the dura mater and the sinuses cut across, a large
quantity of the same kind of blood seen in the heart
will run out. A very limited knowledge of the
brain-box would anticipate such a state to exist in
cholera: *first*, the brain substance, like all other
parts, would be drained as well as all the arachnoid
liquor, and, of course, diminished in volume; *second*,
as this osseous cavity is closed on all sides, excepting where vessels enter and leave, should the brain
diminish, blood would, of a physical necessity, rush
in to fill the vacuum. Hence it is that, in many other
cases than cholera, the sinuses and vessels within the
skull are found to contain more blood than when the
brain has not been diminished in size. The pneuma-
tism of a closed cavity, with unyielding walls, will,
of necessity, suck blood from the nearest sources to
fill the vacuum. This has been fully demonstrated
more than fifty years since. Notwithstanding this
simple physical truth, the presence of this insolite
quantity of blood is everywhere called—

192. *Congestion,* " *The brain is much congested in
cholera,*" is repeatedly stated, as if congestion of a

part constituted a diseased state there. The hackneyed word congestion is much in use, and a very
convenient one to explain to the ignorant something
which he who employs it does not exactly know
himself, and which does not exist, and if present any-
where does not constitute a diseased state. If congestion were disease, scarcely any young woman
would be free from disease of the face, since she often
has even extreme congestion of the face, a congestion that rapidly comes and goes and leaves no trace
behind. Congestion of the sexual organs is, I fancy,
very common in both sexes, and is sometimes much
prolonged and often repeated, but does not disorder
the part in which it exists or produce inflammation,
which is a disease. There is scarcely a single viscus
concerned in nutrition that is not necessarily congested daily, but leaves no trace of disorder. This
word is the creation of a pathologist's imagination,
who, when he cannot on dissection discover the seat
of a disease he is in search of, but finds a cadaveric
appearance due to accumulation of blood, after death,
does not hesitate to exclaim, *great congestion!* Medical writings teem with this convenient word.

193. Having sufficiently discussed the circulation,
or rather the presence of blood in the vessels, let us
now contemplate the blood itself.

194. *Choleraic blood.* In order to comprehend

well what change cholera has exerted on the blood, it will be well to say a word about healthy blood, which varies greatly in the quantity of serum present at different periods, even in the short space of twenty-four hours. When it was the fashion to bleed largely from the arm in pleuritis, some years since, to the amount of 100 to 200 ounces of blood in three or four days, the proportion of serum towards the last bleedings was greatly in excess, at least twice more than what appeared in the blood first drawn. And as regards fibrin, while the quantity appeared great, it was not so, since it was thin and loose. The proportion of red globules was still more diminished. These facts were well known then, and accurately explained thus : the first ten to twenty ounces drawn abstracted equally of the three constituents of the blood, serum, fibrin and globules; each successive abstraction took away equally of what remained; but as fibrin was less rapidly generated than serum, it diminished with each bleeding; and as regards the globules, which require a long time to be created, they were greatly less in number than any of the three constituents, and in certain constitutions are never fully restored. The *volume* of what is called blood in these cases is restored to its maximum each day by the fluids drank by the patient; but this volume is not ordinary blood—it is diluted blood.

195. Now observe what takes place in cholera: the very reverse of what has been just mentioned. The serum is nearly all quickly drained off; the fibrin is decomposed, and scarcely enough remains to hold the globules together; while nearly all the globules remain behind.

196. *Salts in choleraic blood* are rapidly dimin-ished, so much so that in severe or protracted cases scarcely any remain. Free alkali is totally absent, and the neutral salts are mostly absent; but these salts are found abundant in the white matter dis-charged from the bowels.

197. *Inflammation.* From what has been stated it must appear that inflammation cannot exist in any part of a choleraic patient. In the cyanosed state a blister cannot be raised by cantharides, nor by boiling water, moxa or red-hot iron. In mild cases, and on the turn of a case, mustard I have seen act and redden the skin.

198. One memorable instance of the headlong blindness that adheres to certain schools must be mentioned. When cholera invaded western Europe in 1831, inflammation, or its cousin congestion, was still greatly in vogue, and led practitioners to pre-scribe accordingly. One celebrated professor, Del-pèche, deeply tinctured with the doctrine, boldly declared throughout Europe that he had discovered the seat and cause of cholera—it lay in an inflammation of the semilunar ganglion. Being a professor and a great man his assertion was quickly copied into the medical press. He was convinced of his discovery; it needed only bleeding to effect a cure. He went among the different seats of medical learning demonstrating his discovery, and succeeded in convincing *all* those who bow in reverence to a name, until he reached Dublin, where, with a confidence that did honor to his enthusiasm, he would again demonstrate his discovery in the Irish schools. A subject was on the table, he opened the abdomen, searched for the ganglion, showed it in triumph—a red body—being red it was, of course, inflamed. But, lo! what a sudden fall from the pinnacle of greatness to utter insignificance was there when the Dublin anatomist, Harrison, demonstrated that Delpèche's ganglion was simply a lymphatic gland, colored by stasis (congestion); and by a further search he, Harrison, exposed the *real* semilunar ganglion, healthy, and which the professor of Montpellier had so often thought he had demonstrated, but could no longer find. At once the professor fell into deep oblivion, until his tragical end brought him once more before the public in contempt and pity.

199. *The stomach and intestinal canal* are much contracted, dark colored from congestion. The mucous membrane is soft, and more or less covered with that white matter which is characteristic of cholera.

200. The omentum and intestinal adeps are greatly diminished, and both are dark colored from stagnant blood. The parietal and visceral peritonea have lost their clear, shiny appearance, and are quite *dry*.

201. *The liver* does not look much changed, except being darker than usual. The gall bladder is always full of bile, apparently natural, while the hepatic and common ducts are pervious and offer no obstruction to its escape. Here naturally arises the idea, how happens it that the violent and repeated straining to vomit should have spared the gall bladder—for, besides being always found full at death, during life it poured no bile into either the stomach or intestines.

202. Of the *spleen* I have no distinct recollection, beyond remembering that it was small.

203. The *kidneys* suffer no alteration. The *urinary* bladder is always empty, and of the form mentioned in Sec. 135.

204. All the serous membranes, peritoneum, pleura, pericardium, arachnoid, are *dry*. Ascites and hydrothorax, as well as hydrops pericardii, are quickly oozed away. Varicose ulcers, and all other ulcers and sinuses, dry up, but to return, should the patient recover. Two cases of most virulent and copious gonorrhœa ceased at once, but returned on recovery from cholera.

205. The muscles are harder than usual, and dry. On this account, the great diminution of fat, and very slight tendency to decomposition, cholera subjects answer admirably for dissection.

206. *Adipose substance* is greatly diminished in quantity. A patient who has oozed largely by the skin quickly grows lean, the features shrink; in many cases, especially of plump, fat-faced girls, the skin of the cheeks approaches the malæ, the lips become thin, the nose sharp, the eyes deeply set back into the orbits, the corneæ flaccid, and the conjunctiva quite dry. In one short, thick-set lady of fifty years of age, who, before the attack, was remarkable for a protuberant abdomen, due to a great accumulation of adeps between the muscles and skin, her case was severe, but she recovered sufficiently to run into the "second stage," and died on the tenth day, when the heretofore plump belly was greatly reduced, the skin lying in flabby folds. Within the abdomen the omentum is much reduced, the mesentery also, as well as the fat about the kidneys. In naturally lean persons the subcutaneous cellular tissue is tough and less easily cut than in cases of death from other

8

causes than cholera. The heretofore plump hand becomes lean; the skin on the back of the hand lies against the tendons; the fingers become lean and shriveled, like when long soaked in water.

207. The contemplation of this great waste of adeps leads us to approach the subject of *etiology* of cholera, and to inquire what it is that has produced this great change in so short a time as to leave so little fat remaining in some cases; while in all known diseases fat disappears from two causes only: one, want of supply through deficient nourishment; the other, it is removed by absorption—both slow processes, of long duration, while in cholera the removal is effected in a few hours, and without the intervention of absorption and elimination.

208. The itinerary of cholera, and its attack of the healthy only after they have been exposed to one already infected, or to infected materials, clearly prove that the immediate cause is a poison imbibed by the patient, who, in his turn, generates the same kind of poison, capable of infecting others; but what the poison is like cannot be demonstrated; it is known only by its effects. Nor can the contagious principle contained in any known morbid poison be demonstrated; for instance, the pus of syphilis, variola, the exhalation from measles or scarlatina, well known diseases, there is no possibility of showing it, and can

only be known by the effect it produces. We are then forced, in cholera, to infer, as we do in all diseases, that something **has** been introduced into the system which creates the disease. In the case of cholera, after studying the phenomena that the disorder invariably manifests, we are forced to believe that a poison has been introduced into the system, which, totally unlike any poison the product of disease, produces no *disease*, but, instead, a certain catalysis or liquefaction of certain elements of the body, the nature of which liquid so produced has a strong and rapid tendency to reach the cutaneous and intestinal surfaces in a direct way, ooze through the intervening tissues, without the assistance of absorption, circulation and secretion—for all these functions are absolutely suspended in cholera. But the direct transudation does not depend altogether on the quality of the new formed fluid, but does in a great degree depend on an unknown change in the tissues that give it passage, is abundantly proved by the facility with which *saline injections* escape, almost as soon as forced into the veins—escape in the same way as the cholerai matter does.

299. The serum of the blood, and the juices contained in the muscles, rapidly ooze out, but not as serum; the fibrin of the blood is also drained away, but not the globules. The fat is removed, but not in

the quality of fat. Neither serum, nor fibrin, nor fat
can be detected in the escaped matter, which is of a
uniform and special character, quite different from
the materials from which it is derived, and is a new
creation. Nay, more, the fluid of dropsies, such as
ascites, hydrothorax, hydrops pericardii, and, perhaps, ovarian dropsies, also abscess, like milk abscess,
and that of fluctuating bubo, are all converted into
the special choleraic fluid, and as such is discharged
from the system.

210. We must then infer that the choleraic poison,
whatever it may be, has the power of converting certain constituents of the body into a special liquid,
hitherto unknown; also of converting deposits, the
result of disease, into the same. Its power is catalytic over certain tissues and deposits, and the result
is a colliquation.

211. The great strides organic chemistry has
made since half a century comes to our aid, and fur-
nishes us with *analogical* facts illustrative of the mu-
tations referred to.

212. The generation of the choleraic poison, like
that of all known morbid poisons, is due to vital
action within the person. It is more rapidly created
in some constitutions than in others, for we find that
a healthy person, landing in a contaminated place,
has been attacked within twelve hours after expo-

sure; but, in the majority of cases, a period varying from twenty-four to ninety-six hours intervenes before the break-out. This poison, like all morbid poisons, undergoes a period of incubation, as it is called, before it can act on the patient. The same thing occurs in all cases of morbid contamination. Inoculated variola takes nine days' generation in the system before it kindles fever—the subsequent stages of the disease being periods of three days each. When variola is taken spontaneously from the atmospheric contagion there intervenes a period of fifteen to eighteen days of uninterrupted health before the commencement of syaxhus, after which the periods of the stages are similar to those of three days each, like in the inoculated cases. Other poisons, such as scarlatina, measles, etc., have each its period of incubation, during which the health of the patient is not altered. The received contagion lies apparently dormant for a while, long enough to generate the necessary quality and quantity of poison (lately called ferment or zymosis) to set up the same disease in a new patient. An analogous process is comparatively carried on in cholera, but here there are no periods of definite duration; the only one it has is of colliquation, arrest of circulation and aeration.

213. We are forced to believe that this poison,

like those already mentioned, and many others, does not manifest its power until it has acquired a certain quality, also quantity; for, if only partially created, like in case of ordinary ferments which metamorphose vegetable bodies, it may prove defective in quality and consequently inefficient, or it may be insufficient in quantity to overcome the resistance that some constitutions offer to its virulence. We may also suppose that in those cases which run a rapid course, terminate suddenly and leave the patient restored to health, only weak for a day or two after, that the poison was insufficient to colliquate the whole system, like in bad cases. While in other cases the poison will be sufficient to saturate, as it were, the system, and act until death results; or, that the idiosyncracy of the patient will resist, and an end to the storm is attained, leaving an exhausted economy, with a flickering *vis medicatrix naturæ*, which attempts to make a cure, or not, and run into "the typhoid state."

214. We may suppose or imagine that the period of incubation and complete generation of the poison was a diseased state of the system for the time being; if so, it is the only period in which disease exists, though it is never manifest. But the moment the poison operates, it causes a complete cessation of nearly all the usual physiological actions—circula-

tion, aeration, calorification, digestion and its se-
quents—repair and elimination—are all completely
arrested, while the cerebral function remains undisturbed, independently of the cessations just named.

215. The action of the poison establishes a colliquation by means of a hitherto unknown catalytic
cause, in the total absence of any recognizable disease. Like all other morbid poisons, it selects only
a certain number of persons out of the whole population, and this without the slightest regard to the
state of health the person is in at the time of exposure, as is made manifest by its attacking persons ill
of fever at the various stages of the fever; regardless
of the dyscrasic taints, such as syphilis, scrofula,
phthisis, even persons under mercurial salivation;
regardless of localities reputed to be healthy, cleanly,
or otherwise, refuting the ceaseless and senseless cry
against dirty and filthy places; regardless of the trade
or calling of individuals; regardless of good or bad
habits—among the latter drunkards, who, by-the-by,
are proved by statistics to be more exempt than well
conducted persons.

216. Hitherto I have avoided speculation, called
by some, theory, by others, hypotheses; but now, for
once, will venture to offer what follows, not because
I believe it is true, but because it may set some one
thinking how the poison of cholera can operate.

Any one familiar with that part of organic chemistry known under the name of Compound Elements, will readily conceive how a poison can be generated in the system, and being once formed in suitable quantity may exert a catalytic power, by its mere presence, over tissues which are in their natural state, and convert them from semi-solids into a liquid that shall ooze through the cutaneous and intestinal surfaces, but not through serous membranes or through any of the other mucous surfaces, excepting the intestinal.

217. Every one knows that a compound radical may be changed into another one, consisting of the same precise elements, and even of the same number of atoms, by the mere presence of a quite different body, which imparts nothing and abstracts nothing from the body it alters. The only thing it does is to alter the grouping of the old body to a new series of grouping in the new body its presence has brought about. Alcohol is converted into ether by the *presence* of sulphuric acid; not a particle of the acid is given to either the alcohol or to the ether. Sulphuric acid, by its mere presence, converts starch into gum first, subsequently the gum into sugar, without combining with either, or losing a particle of itself. The same conversion of organic matter is effected by various ferments. Many other examples of the power

of *presence* are well known, and that produced by platinum is notorious.

Knowing these facts, we cannot be surprised to hear it stated that the presence of the choleraic poison will rapidly convert the living juices and semi-solids of the body into a peculiar fluid, hitherto unknown, which has the property of transuding to and through the surfaces.

218. Some writers have asserted that the gruel-like matter discharged was the fibrin of the blood; but the quantity discharged is greater than the amount of fibrin contained in the blood. On what imaginary ground this has been stated no one knows, nor can it be proved. No doubt fibrin, as well as other elements and parts of tissues, is converted into the new liquid, but there is no chemical operation that can prove the matter in question to be fibrin. It has none of the properties of fibrin, and all that chemistry can do is to resolve it into elements which are common to all animal matter, healthy or otherwise, consisting of C. H. O. N.

219. To sum up what has been stated in this chapter, it appears that the human cosmogony, proportioned to the size of the system, has suffered a cataclysm, the greatest known. A portion of the constituent tissues are suddenly converted into a new fluid, an alarming colliquation is the result; the new

8*

liquid resists the confinement presented by the
envelopes of the body, makes its way to the surface, escapes in profuse quantity, drains the body
to dryness, and, in over one-third of the cases, to
death. It is a colliquation not only of the living
constituents of the body, but even of collections of
matter that no longer form part of the system—such
as dropsies—which are carried away out of their re-
ceptacles along with the choleraic avalanche. During
all this commotion the intellectual system sits serene,
conscious of the revolution, but helpless. The newly
created fluid has no taste to the patient, nor did it
impart any to a few enthusiasts who ventured to
swallow it; it is destitute of odor, although a choleraic
patient produces in his room an indescribable faint
smell, resembling that exhaled from water in which
spinage has been boiled.

CHAPTER VII.

220. A catalogue of remedies which have been used in cholera, asserted to have cured nearly every patient:

1. *Bleeding*, cupping, leeching, arteriotomy, have cured, in the practice of one physician, eighty cases out of eighty-two. Infallible.

2. Calomel, with or without opium, in grain doses every half hour—in four, ten, scruple, half drachm, drachm and two drachm doses, every half hour, in the practice of certain celebrities. Infallible.

3. Emetics of mustard, ipecacuanha, emetic tartar in five grain doses, kitchen salt in water, frequently. All very beneficial.

4. Pressure over the liver. Surely arrests vomiting.

5. Purgatives of castor oil, Glauber's salts, Croton oil, aloes. Highly beneficial.

6. Sedatives, as prussic acid, laurel water. Charming.

7. Enemas of hot water, spirits, asafœtida, tobacco,

spirit of turpentine, of brandy and water. Very beneficial.

8. Stimulants. Brandy, sulphuric ether, carbonate of ammonia, creosote, strychnine, phosphorus, cacaorum distilled over horse-dung, camphor, capsicum, horse-radish, garlic. Many cures.

9. External heat. Warm baths, hot oats, hot sand, hot ashes, a warm skin fresh from a sheep. Did great good.

10. Ice on the spine and back. Excellent.

11. Astringents. Sulphate of copper, acetate of lead, nitrate of silver in pills, muriated tincture of iron, lime water with milk, extract of catechu. All beneficial.

12. Drinks of brandy punch. Cured many.

13. Acupuncture of the heart. Uncertain.

14. Galvanism. Useless.

15. Quinine cured like in ague. Good.

16. Charcoal. A sovereign remedy.

17. Subnitrate of bismuth. Phosphorus.

18. External or revulsive applications. Scalding water, cantharides, strong nitric acid over the stomach, moxas, red-hot iron, heated hammers and laundry irons rubbed up and down the back. Produced great effect.

19. Laughing gas inhaled, oxygen gas inhaled, also drank in impregnated water. Useless.

20. Saline injections into the veins.

21. Opium variously combined, etc., etc., etc.

The foregoing list of heterogeneous remedies is not the production of quacks, but was seriously published, strongly recommended by practitioners of eminence, and who, to say the least, ought to have known better than to publish their conceits, and ought now to blush at their errors.

221. If ever humanity were entitled to pity for sufferings created by errors, the growth of a false science, it is surely in the case of cholera, as a glance at the remedies and treatment employed mentioned in the foregoing list would indicate. Very few practitioners have had the conscience to avow that they were at a loss what to do, and had also the courage to resist the importunities of interested persons urging them to " do something." A majority of practitioners went to work in a case of cholera trying one thing or any thing; and should the patient survive both the pest and the remedy, they would hasten to publish a cure effected by an infallible mode of treatment. After repeated failures, they would jump to another means as senseless as the first one; and now, forgetting the late asserted cures, proclaim the last one infallible also! In this way has the profession been disgraced.

222. It is easy to perceive, through the mist rising
from the catalogue of remedies just cited, where lay
the root of a false medication. The mind of a student
receives a bend suited to nosology, the symptoms and
treatment as laid down in books, and which he finds
confirmed by a most imperfect, coarse and erroneous
physiology, grown out of a mixture of descriptive
anatomy and shallow chemistry. Thus endowed he
feels himself ready to combat any disorder, and that
he is armed with a knowledge of every thing per-
taining to medicine—known and unknown. He
never once notices the conflict of dogmas that ceaselessly rise up in his path, unless it is to reconcile or
refute them, always to his own satisfaction, and
never to that of a co-laborer. He has the faculty of
argumentation. When he is at a loss for a method
of treatment, he soon finds one, founded on analogies,
whether the supposed analogy suits or not. Hence
the wild range of his fancy, to which he bows with
complacent respect, adapts something to suit something, right or wrong; who is to judge? Not the
patient, or if he or others do, his acquired facility
of argumentation and explanation will carry him
through in triumph, to his own satisfaction, if not to
that of others.

Thus armed, he attacks cholera as he would disease
as described in nosology, not once dreaming that

nosological disease does not exist in cholera. He brings a treatment founded on analogies, but which do not exist in reality in cholera. His patient vomits—he administers carminatives, to soothe a stomach that is not disordered in itself; his patient has profuse purging, when the purging treated of in nosology does not exist—he adminsters astringents; his patient transpires profusely—this he dare not attack, for now he is confused; the heart and pulse are feeble, the latter extinct—he prescribes stimulants to organs that cannot be roused by stimulants; his patient is cold—he will apply external heat that cannot penetrate to the interior; the stomach is still irritable—he will apply revulsives where metastasis cannot take place; he will excite a " *new action in the system,*" founded on the theory that two diseases cannot exist at the same time; he gives calomel for this purpose, a substance which is perfectly inert in the actual and wonderful state of the system.

223. Let us now do as was done with the symptoms, examine them in detail, and the treatment and remedies, each by itself, and discover, if we can, the discrepancies we shall meet with.

Abstraction of blood. When contagious cholera broke out in India, the prevalent idea of the nature of the disease was, if not inflammatory, there was at least congestion somewhere. At that time bleed-

ing was the battle-horse, and often had recourse
to, if only for fashion's sake. Accordingly we find,
in the first years of the pestilence, almost every practitioner drawing blood, urging the practice in publi-
cations, and citing numerous cures. They advised to
bleed to the extent of thirty ounces, if that quantity
can be had; if not, get as much as you can; and as
the blood soon becomes thick and flows with reluc-
tance, open the vein as early as possible, while the
heart has still power to act, and the blood not too much
inspissated to flow. If phlebotomy gives no blood,
apply cups; if cups give none, try leeches; if all these
fail, try arteriotomy. Some practitioners declared
that under the sanguinary treatment they *cured* nine
out of ten cases; others cured nineteen out of twenty;
and Dr. Barrel eighty out of eighty-two cases, all by
bleeding! Other practitioners were less fortunate,
soon saw the error and opposed it; but still the practice prevailed for several years, even until the pest
reached Europe in 1830, and after, by those routinists
who can never get out of a beaten track.

224. We find, from the very beginning of an attack
of cholera, that all venous capillary action is greatly
arrested, if not quite stopped, and that the vigor of
the heart diminishes rapidly. The volume of blood
is soon reduced by the oozing out of the serum and
much of the fibrin, leaving the globules behind. Can

bleeding arrest or restore the loss, or re-establish the natural proportions of serum, salts, fibrin, and the quality of the globules? No. Besides, blood, like the tissues, is subject to the colliquative force of the poison, and is itself inert, receives nothing, communicates nothing, and of course is a passive body. Hence it is that inflammation has never been excited, and cannot be kindled by any amount of irritation during the choleraic stage. Will bleeding, then, arrest an inflammation that cannot and does not exist? Absurd. Experience has at last forced the advocates of bleeding, and the routine practitioners, to abandon this gross error.

225. *Calomel* is the next heroic remedy, which, in India, appears to have been, as it were, a household necessity, to be administered in every and all kinds of disease. Given as a mere laxative, as a depurative of bile and other humors, or with a higher view, under the theory of inducing a change in the constitution for a time by establishing a new disease, which is to supplant the first one and expel it, under the dictum that "two diseased actions cannot exist together." Leaving the theory to defend itself, let us endeavor to ascertain whether there exists the remotest or slightest probability that the drug can be absorbed into the system; if not, it must prove useless. When practitioners had recovered from the alarm and confusion of ideas, and commenced slowly to observe the phe-

nomena produced by the poison, it was discovered
that absorption was completely arrested; consequently, calomel might be conveyed into the stomach, but
could not enter the system. Besides, it was moreover
discovered that circulation was also arrested; so that,
if even mercury were absorbed, it could not be carried through the system. So potent were a wrong
education, a defective physiology, a badly acquired
habit, a blind faith in false doctrines, and a perni-
cious obstinacy in adhering to habit, that almost all
physicians made use of calomel. One gave a grain,
with or without opium, every half hour; another two
to five grains; finding these doses inefficient, ten to
twenty grains; others, bolder, gave half drachm, one
and two drachm doses. Should one or two patients
out of three (the average number) recover, the recov-
eries were boasted of as *cures*. For a long time the
small dose, the medium, and the heroic dose doctors
published how successful was their practice. In time,
however, this abuse of calomel declined, but is not
yet extinct among that numerous class of practitioners who cannot rise above the grade of routinists or
mere medicasters.

226. *Metastatic treatment.* The theory of metastasis, an ancient one, not altogether false, is still
prevalent, and comes as a kind, too often a cruel auxiliary to the practitioner who, when at his wit's end

for means, calls in the aid of his friends—revulsives—
in the hope that they will translate a malady to the
surface, if not quite, yet in some proportion. To
leave nothing undone, even when nothing can be
done, sinapasms, cantharides, cauteries, and every
imaginable means capable of irritating and injuring
the surface, have been employed without stint. When
employed early, which is rarely the case—that is,
before the surface has become quite cold and the
capillaries stagnant, as announced by cyanosis—these
means may act slightly on the skin, for a very short
time, should the case be a slow one; soon, however, to
be arrested, as all other actions are, by the cataclysm
induced by the poison. But in most cases the length
of time claimed by all epispastics to act is longer than
the collapse will permit, and vesication, if useful, cannot take place. Why, then, torment the patient with
them ?

227. A more energetic metastatic means has been
had recourse to—scalding water, as an instantaneous
means; red-hot iron, by French practitioners indoc-
trinated to this cruelty by the combustible old Larrey;
also his moxas; by the Germans, such delicate instru-
ments as burning-hot hammers and laundry irons
promenaded down each side of the spine and back;
strong nitric acid applied over the stomach ! Was I
wrong when I said the choleraic were entitled to

pity for the sufferings inflicted by dogmatic practitioners? It is fortunate for the victims of such monstrous cruelties that none recovered to endure the protracted suffering that wide-spread sloughing would induce. If any of these patients recovered it is unknown, and the heroic doctors have kept silent regarding the ultimate destruction that their barbarity must have created, had it been possible for man to resist the first effects. Yet many of the medical heroes had the hardihood to declare that their treatment was good, and did cure.

228. *Emetics to arrest vomiting.* The signal benefit of an emetic in cases of saburra, and at the commencement of fevers, suggested, no doubt, the idea of " clearing out the stomach," and in this way remove peccant matter, which caused the vomiting. Had the majority of remedies administered in cholera been as reasonably founded on derived experience, no great censure could attach to the practice; but, unfortunately, the peccant matter is renewed as fast as it is ejected, and as long as the system can furnish material for colliquation; but the case bears no analogy to those in which emetics are useful.

Tartrate of antimony was one of the forms of emetic employed and praised, given in five grain doses, repeated. Such a dose is astounding, and finds no place in practice, excepting in the Italian

practice wherein it was so administered, not to clear
the stomach, for such a dose does not excite vomiting, but to produce such an arrest of circulation and
nervation (the new word is innervation) as will subdue pleuritic inflammation, against which it was employed with benefit. But cholera is neither pleuritis
nor other inflammation, nor a disease, as has already
been explained. Ipecacuanha, mustard, and kitchen
salt, frequently repeated, although an erroneous practice, is not deleterious and astounding like antimony.

229. *Pressure over the liver*, to arrest vomiting,
was recommended, with what effect is not mentioned,
nor on what principle is it possible to imagine—
although in this instance, like in many others, there
are metaphysical abstractionists who are ever ready
to explain any thing, if it be only absurd enough.

230. *Purgatives* were had recourse to and recommended. They could do no other harm than what is
due to the annoyance of swallowing them and their
immediate rejection upward, adding to the distress
of the patient and increasing his exhaustion. But,
were it possible for the purgative to lie on the stomach and reach the intestinal canal, it could not come
in contact with the mucous membrane, which is in a
ceaseless state of oozing, a state that will protect its
surface from the remedy; besides that, as all physiological action is here arrested, a purgative, even

could it reach the mucous surface, would not excite a purgative secretion.

Castor oil, aloes, Glauber's salts, Croton oil, ox gall, were the purgatives employed.

231. *Antispasmodics* were called into service—on what ground no rational mind can conceive; but the practice can be defended and explained by routinists, who have the faculty of explaining things known and unknown, right or wrong. Among these remedies are, or were, musk, cajeput oil, guaicum, asafœtida, hartshorn, valerian, or any thing that has a bad odor.

232. *Sedatives*—Prussic acid, laurel water, opium; but of this last see hereafter.

233. *Enemas*—not able to do good *per oram,* some practitioners descended *ad infra* to operate *per anum.* Accordingly, enemas containing asafœtida, castoreum, spirit of turpentine, alcoholic spirits, even tobacco, have been recommended and used; and fortunately the tobacco has been rejected as soon as injected, else it would add a new collapse to that which the patient can no longer bear up against.

234. *Stimulants* have been largely had recourse to, and might, by judicious management, prove an auxiliary to other means; but they require a most careful watching. Sulphuric ether alone, and with ammonia, carbonate of ammonia in five grain doses, creosote, strychnine have been given; phosphorus also—here I must stop a moment to say what I have seen done with phosphorus. It was recommended in the Edinburgh *Journal* to give pills of phosphorus, made by beating phosphorus and bread-crumb under water into a mass and making it into pills. An old gentleman caught the idea, made some and gave them. His son, a dashing new light, disdained trammels, saw no necessity for following the laborious method of his father, but attained the same end by simply cutting sticks of common phosphorus into chunks. He carried these pieces with him in a vial in water. As I was passing a house in a garden, a woman standing at the gate asked me to walk in and see her husband, Wm. Harrison, sexton to the English burying ground. I found him in collapse, quite sensible. The young doctor followed me in, who, without ceremony, said, "My dear fellow, I'll cure you at once. Open your mouth; I'll poke these pills down your throat—they'll cure you." Accordingly he took out of his bottle about *ten* pieces of phosphorus of at least two grains each, which he did "poke" into the pharynx. How the remedy acted I did not stay to see; but he died a few hours after. This man had gone unscathed through the whole of 1832, and was daily at work conveying dead to the ground and burying them. He passed through the greater part of the cholera of 1834, but was attacked August 7th, died the same day, aged 54.

Opium with camphor, opium in *large* doses, recommended by Dr. Tweedie, who ought to have known better, and who at last said while it did good it hastened death.   Camphor, capsicum, horse-radish; brandy alone will cure, said Dr. Leo, of Vienna ; especially cacao-brandy, distilled over horse-dung, called Tangara.   Dr. Kiven's infallible Vienna remedy, which cured all but two out of 242 cases, was composed as follows : one pint of spirits, half a pint of vinegar, one ounce of camphor, two drachms of pepper and one ounce of garlic—a teaspoonful to be taken frequently.

Practitioners who have a predilection for stimulant medicines of a carminative class, such as sulphuric ether, might, perhaps, substitute chloroform in five to ten minim doses in a little water, internally, since it is agreeable to take, and as good a stimulant as ether. But not by inhalation; for, should it permeate the lung-tissue, which, in cholera, is not likely, it would tend to paralyze the heart in an additional degree. Besides, chloroform in the stomach, like carbonic acid gas, produces the opposite effect to what occurs in inhalation.

235. *Warm baths*, also dry heat afforded by hot oats, hot sand, hot chalk in powder, hot ashes, and, as

a contrast of consistency, ice on the spine, and cold
baths.

236. *Astringents.* Sulphate of copper, by no less
a character than the great Dupuytren, nitrate of silver in pills, acetate of lead, muriated tincture of iron,
lime water and milk, magnesia, subnitrate of bismuth
in four grain doses, extract of catechu.

237. *Drinks.* Warm water, cold water, *ad libi-
tum;* by others, none at all; brandy, pure or as punch;
effervescing draughts.

238. Acupuncture of the heart, by Dr. Searl, at
Warsaw, one of the commission sent thither to inquire into the pest and report upon the best treatment. He was, of course, a respectable practitioner,
but this conduct proved him to be a crazy man; by
the same, galvanism; he, however, says, correctly, that
this last is of no service.

239. *Quinine, per oram et anum,* has been administered by routine practitioners, who had not
strength to move out of the beaten track of nosology,
and who compared an attack of cholera to one of
ague, supposed that the collapsed state was the cold
stage, and the weak attempt of nature to recuperate,
called the typhoid state, was the hot stage. What
can be done with such minds? Quinine was also given
in enemas when it could not be retained, and even
injected into the veins.

9

240. *Charcoal* made from burnt corks is reported to have been used on board of a United States frigate in India, most likely as an antiseptic. This idea was caught up by an extraordinary man, poor, philanthropic, ignorant and zealous. He appears suddenly in Montreal, soon after the outbreak, accompanied with his inseparable companions—a lean mare and two grown-up colts; these followed him wherever he went, like docile dogs. His services were gratuitous, kind and modest. In a short time he was looked upon by the poor as a saint from heaven, and got the name of Saint Roch. His remedy was equal parts of powdered wood charcoal, hogs' lard and maple sugar. Of this disgusting medley was to be taken as much as could be swallowed; an addition was made, consisting of a pailful of cold water and wood ashes, with which the feet and legs were bathed. This lye he rubbed, or caused to be diligently rubbed into the skin, as high up as to the knees. I saw him at work in the case of a tall, lean old man, whose skin over the tibia appeared to be dissolved and rubbed away. This man was a wealthy and respectable publican, George Wurtele, about sixty years of age. He sat in a chair during the treatment, quite conscious, and died the same day, after a few hours' illness.

241. *Oxygen gas* was had recourse to, suggested by the absence of aeration of the blood. It was both inhaled and drank in water saturated with it, also in saturated water injected into the rectum, but proving of no avail, was soon abandoned. Nitrous oxyde gas was used in the same way with the same results.

242. *In* 1848 *chloroform* was administered, internally, with beneficial effect, I was told; but of myself I have no knowledge of its use.

243. *Saline venous injections.* The able and graphic description of a case of cholera by Dr. O'Shaughnessy, and his analysis of the blood and discharges, led to his first suggestion of the injection of a solution of nitrate or chlorate of potash, as being highly oxygenated salts and likely to act on the globules; but it does not appear whether or not it was put into practice. Dr. Clanny, of Sunderland, also threw considerable light on the chemistry of cholera. But the first one to avail himself of injecting salts into the veins was Dr. Latta, of Leith, Scotland. His solution was composed of chloride of soda, two drachms, carbonate of soda, two scruples, white of egg, and water, four pints. He injected in the beginning of his operations from five to ten pounds, repeated at short intervals; but seeing the impunity with which this foreign matter could be introduced into the vessels, he grew bolder, until he and many others injected, without stopping, 330 ounces in the space of **twelve hours.** Thirty-one pints were injected into a patient in the space of three hours. As much as six hundred ounces have been injected at a temperature of 112°, 116°, 124°, but oozed out cold *per cutem et anum*. However, it was soon ascertained that it gave no permanent benefit.

244. The effect of the injection was of the most marvelous character, and often struck non-professional beholders with awe. A patient lying in the collapsed state, lean and cold, and often manifesting no consciousness, moribund, would, as the injecting progressed, gradually fill out, improve in color, open and shut the lids, and sometimes speak. Dr. Tweedie mentions three wonderful cases, which, after being resuscitated, as it were, got up, conversed jocularly, but all at once fell into collapse again and suddenly died.

245. Transfusion was in some cases added after the saline injection, but no good came of it.

The idea of injecting salines was with a view to oxygenate the blood, and on that account O'Shaughnessy suggested highly oxygenated neutral salts; but Latta employed carbonated salts, and others followed his example—yet the effect was marvelous. Soda or any neutral salt will redden the blood.

246. *Personal experience.* I shall cite only two cases out of many. A chambermaid, a beautiful girl, much admired for her plumpness and fine color, about twenty years of age, lived at the Brock Hotel, Montreal. While in the cyanosed and collapsed state, lean and shriveled, especially the fingers and hands, but possessed of her senses, Dr. Stephenson injected into the basilic vein of one arm five or six pints of saline fluid ; gradually her face filled out, her cheeks became once more rosy, her dry and sunken eyes became moist and bright, and less deeply sunken into the orbits; her voice became less husky—altogether she seemed restored, but a shivering came over her and she soon died. This was the first case I saw.

247. Another case. I was invited to see Drs. Arnoldi, Senior and Junior, inject Mr. Oliver Wait, on Sunday, at noon, August 21st, 1832. He was about forty-five years of age, a very strong and muscular man, much given to daily active exercise in the open air. I found him lying on his back, apparently dead, excepting the quiet respiration manifested by the slow movement of his chest. Vomiting, purging, and oozing out by the skin had completely ceased; his skin was cold, leaden color, but that of the face, tanned by habitual exposure, was rather brown than blue; his face was now sharp and lean; his eyes deep sunken in the orbits ; the lids partly open and immovable; the eye was quite dry and the cornea flattened; his mouth was slightly open; house flies crawled in and out of the lips, into the nostrils and out, and over the exposed part of the eyes, without exciting the least perceptible winking. Such was his state when the old doctor had finished making two large wash-bowls of saline fluid, and his son had inserted the tube into the basilic vein of his left arm. All being ready, the old gentleman said to me, " You do not countenance this treatment; here is an extreme case; you shall soon see him restored." He commenced to inject by means of a Reed's pocket brass syringe, capable of discharging about twelve drachms at a stroke of the piston. He worked on deliberately and ceaselessly until the first bowlful was nearly used before any apparent effect became noticeable; at last the color of the cheeks became lighter, the face looked a little fuller, and the malæ less prominent; next, the eyes became moist, and much less dim, but the lids did not move ; his respiration was regular as heretofore, but much more expanded. At the last of the second bowlful the lids opened completely; he stared at us, but did not wink, nor did he make the least motion, nor attempt to speak; his skin grew moist, but remained cold, although the injection was at 112°; and at last we heard a gush from the bowels, soon followed by an exceedingly strong rigor. The young doctor exclaimed: " We have established the ague stage; send for quinine !" In a few moments more he ceased to breathe, and was dead.

The moribund state of this strong man before the injection and his sudden apparent recovery, gave to many of the bystanders the idea of a ghostly resurrection of a corpse; some of them left the room in terror. In me it excited the sentiment of a total refutation of the doctrines of physiology—a dead man brought back to life, as it were, by merely filling the vessels with a fluid not natural to the body. The heart was once more set agoing, and a feeble pulse could be felt at the wrist; the mind once more restored—some thought. Can a recently dead man be restored to life —at least, will these experiments be improved upon and made available in disease?

248. Having completed a critical review, I hope with candor and honesty, of most of the remedies and modes of treatment recommended in cholera up to the period this sketch treats of, that is, to 1834, I have only to add one advice—look at the heterogeneous kinds of incompatibles in the quoted catalogue, many of them still in use, each opposed to each, and nearly all useless, often injurious!

It is with a faint pleasure I come to notice and remark upon another class of remedies that do not revolt our common sense, and which have proved to be great auxiliaries to nature—in many cases have contributed a fair share towards the recovery of some patients, and, when judiciously employed, injurious to none.

249. *Opium*, first of all. Sydenham said: "Without opium, all medication would be imperfect and insufficient." Its effect is often decisive of life or death; used at the critical moment it will save, and as surely will destroy if wrongly applied; is unique in its operation, and cannot be replaced by any other remedy; has reigned for more than a thousand years despot over the medical world, and spread benefits as well as ruin over humanity; it is a double-edged sword, a divine gift in the hands of a master, a poison in those of a mere routinist—a medicaster—a demi-physician. *Sacra vitæ anchora, circumspecte agentibus, est opium; cymba vero Charontis in manu imperiti.*"

Throughout the whole reign of cholera opium has been had recourse to, with benefit when judiciously employed; but which, alas, has seldom been the case, for there are only a few practitioners who are deeply versed in the recondite action of this heroic remedy while the great majority see in it only an every-day drug, and are totally ignorant of its mysterious power. This is a severe judgment, but, unfortunately, a true one. In describing common cholera, Secs. 92 and 96,

97, 98, *passim*, I there endeavored to explain how opium, only *one* grain, was sufficient to arrest that complaint.

But first let me make one admonition on the preparation. When a practitioner has serious work on hand, let him use the simplest means as being the most sure to reach the end he has in view. In all serious cases where real work is to be done, employ pure opium. Nature has made it perfect. She is the greatest and wisest chemist, with whom no one can compete. Pure opium contains a well balanced mixture of homogeneous ingredients, the removal of one of which will proportionately deteriorate the article. Pharmaceutists have tortured opium in all possible ways to improve on that which comes perfect from the hand of nature, and by their scandalous puffing have imposed on weak minds, ever on the *qui-vive* for something new; in cases of cholera, abandon, then, all fancy forms of opium—black drop, elixirs, and the various salts of morphine, codeine, narcotine—all the *'ines.*

250. In Asiatic cholera, a small dose of opium, not over *one* grain, given after vomiting has continued long enough to reject the new and noxious matter poured into the stomach, will calm the irritability of the organ, and save the patient from the debilitating efforts at vomiting, no longer needed. A larger dose might so paralyze the stomach as to arrest its power

9*

of ejecting what ought to be discharged, and lock it in, to the injury of the patient. Should the dose be rejected, take a moment of repose and give another, dry, if possible, so that the volume may not offend a preternaturally excitable organ. Should this dose be retained, in fifteen or twenty minutes after some agreeable acid drink, mentioned hereafter, may be given, an ounce at one time; this being retained, repeat it, and increase the quantity and the frequency of repetition compatible with the powers of retention. To give oftener or more at a time before the stomach will retain it will frustrate the object sought.

The real use of opium here is simply to quiet the excessive irritability of the stomach, and arrest the frequent vomiting, the efforts of which tend greatly to add to the exhaustion of the patient. Opium has been combined with aromatics, stimulants, such as ether, ammonia, etc. It is doubtless good, but the *volume* of the dose is greater than the viscus will endure, and will be rejected. Give the opium dry.

Opium has been given in repeated doses, every half hour, regardless of a stomach filled with poisonous matter which vomiting must discharge, and having succeeded in paralyzing the organ, has tended to a speedy death. Before deciding on the administration of opium, one important consideration must be carefully borne in mind, that is, what is the cause of the vomiting? Let us examine this point. It is certain that previous to the attack the stomach was in no way disordered; therefore, something has happened to disturb it; what is this something? Evidently, newly created choleraic matter poured, or rather oozed, into the stomach, which offends it, and, like all offensive matter in the stomach, vomiting is the remedy nature employs to rid the organ of what is noxious. As this matter cannot all pass the pylorus into the intestines it will be rejected, and must be got rid of, to a great extent, before it will be prudent to give opium and thereby arrest the rejective power of the organ. In the early period of the colliquation the noxious matter is rapidly oozed into the stomach, but after a while there is less and less of it so furnished, while the organ once offended continues to act when none or little remains; it will act when all choleraic matter has ceased to be present; if only water be admitted in a quantity greater than the induced irritability will endure, this will be rejected. It is this persistent irritability that opium is employed to allay.

When opium is administered at the proper time, and in proper quantity, it will not only allay a state of vomiting, which is no longer needed, but it will also soothe the whole economy in a notable degree, and pave the way to a recovery. Would it be safe to inject the opiate hypodermically in cases of such great irritability of the stomach that much difficulty to retain the medicine on the stomach is experienced? But it must be borne in mind that the object is to arrest useless vomiting; therefore, the quantity injected ought to be very small.

251. *Water*, cold water, is the ceaseless cry of the patient from almost the first onset of the case. This thirst is not due to fever, or any thing like fever, for fever does not and cannot exist in cholera; it is due to two causes—a perverted sensation of heat, like that of the skin, and to the great drain which has dried up the tissues, as it were. During the early period of the complaint, while vomiting is active, produced by the presence of the new matter, let the patient take as much as he calls for; it will be rejected, of course, but will not increase the vomiting. However, when the colliquation becomes diminished or exhausted, care must be had in mind that now the vomiting is continued from induced irritability after the exciting cause has ceased to act.

252. *Drinks* are loudly called for, and might be given under the conditions already mentioned, were it not that they are useless in the early period of the complaint, since they cannot be absorbed or circulated, as elsewhere noticed. There is, however, one beverage—doubtless there are others—which I have found grateful and useful. It may be called a lemonade, made of sweetened water acidulated with *tartaric* acid, in the proportion of half a drachm to a tumblerful. But it must be given with much judgment, so as not to offend the stomach by volume. I had previously tried nitric acid, but it proved offensive. When phosphorus was suggested as needed by the system, I bethought myself of substituting phosphoric acid for tartaric acid, but being satisfied with the latter, I did not put the phosphoric idea into execution.

253. *Phosphorus* has been prescribed as stated, Sec. 234—upon what principle I cannot imagine. The common stick phosphorus is exceedingly unmanageable, and on this account I would recommend to those who believe in its utility to try amorphous phosphorus, which is quite manageable and pleasant to take.

254. *Hot applications* have already been mentioned, and are useless, besides being an annoyance to the patient. Experience shows that a patient cannot be warmed by external means, and the large amount of hot saline injections into the veins has not contributed to the warmth of the patient. Aeration of the red part of the blood is the means employed by nature.

255. *Cold applications.* I knew one case of cold bathing. A poor man had just lost his wife from cholera, when he himself was attacked, and, like all others, suffered from a feeling of heat; he found his way into the yard, got into his water-barrel, remained in it some twelve hours, when, feeling relieved, he got back into the house nearly recovered, when he found his dead wife where he had left her.

256. The foregoing analysis of the various remedies and modes of treatment employed in cholera show clearly that the authors were at a loss what to do, and quite bewildered; and that no correct opinion existed regarding the immediate cause, nor is that uncertainty yet removed. Let us turn and be honest, cease our pretensions to a knowledge that we do not possess, discard the foregoing errors, and by so doing disembarras ourselves of a false, useless and oftentimes injurious medication, and commence anew. Let us throw theory overboard as useless now, and try instead a reasonable, empirical and an expectant treatment, until, little by little, we arrive at something that can be relied upon. In this way we shall do no harm, and not hurry the patient to his end, as has too often been done. We still meet with practitioners who assert boldly that cholera is quite manageable, and boast of the many cures they have made.

257. *Dialogue.* You deceive yourself, Doctor; neither you nor any one else has ever cured a case of cholera.

Yes, I have cured many; I can prove what I say. There is Mr. — and Mrs. —, and many others, who were very ill; I cured them, and they are now alive, and will substantiate what I say.

No doubt you had such patients; that you attended them, and that they are alive to day; so far true: but how many cases have you had, and how many of them have survived?

I have cured nearly every case—at least nineteen out of twenty.

It is likely that you trust more to memory than to a correct list of your cases and alleged cures. All history, on which the slightest reliance can be placed, and respect accorded, shows that the most favorable result has been the recovery of two in three cases; one-third of the attacks have died. There are instances of one-half, even of two-thirds, of the attacks ending fatally in a few places invaded. If you can be persuaded of your error, you will, on a future occasion, judge with more accuracy than you now do, and be in a better state of mind to benefit a patient than heretofore.

258. *Expectant treatment aided by an enlightened empiricism.* Before deciding on a method of treat-

ing a case of cholera, we must make up our mind
regarding the nature of the complaint—that is, the
immediate mode in which the poison acts. It has
been shown (Sec. 210) that this mode of action is
one that converts a portion of the body into a liquid
form, and that the liquid so created will ooze through
the tissues and out of the body by way of the intestinal, mucous and dermoid surfaces. Whether the
liquid which escapes existed in the tissues already
formed before it reached the surfaces, or that it put
on the liquid form on arriving at the surfaces, is uncertain; but in any case, having arrived there, and been
poured into the alimentary canal, it must not be
locked up therein by any medication we may employ,
should we possess a medicine capable of arresting its
discharge. There are well authenticated cases in
which no discharges have taken place, unless it is that
which escapes by the skin. In such cases it is evident that the colliquated matter has not entered the
stomach and intestines, but is retained within the
tissues; perhaps it did not go through the necessary
phases that precede the liquid form. However this
may be, it is of little importance, in a medical point
of view, since all such cases are mortal and cannot
be helped by medication.

259. A practitioner seldom sees the commencement of a case, unless he should happen to be near

the person attacked. When such an opportunity occurs, the first complaint of the patient is that he feels an indescribable feeling of general uneasiness— something wrong at the stomach—a mawkish sensation, and a rapidly increasing debility. He has a call to stool, which he obeys; a large motion escapes, unaccompanied by any pain in the bowels or the least feeling of colic. He has scarcely relieved the bowels when he feels another call and has another motion, succeeded by another, and so on. He now begins to experience some thirst and a pasty mouth, destitute of taste, soon followed by vomiting; a damp skin, which is cold to another, but not to himself, and which he soon complains of as being hot. Such were the feelings of the author when he was attacked at a time when he was quite well, in the forenoon. In general a doctor does not see his case until it has advanced to rapid purging, vomiting, cyanosis and cold surface.

260. Meeting with a case in this stage, the first prescription must be to inculcate as much quietude, or cessation of that tossing about from side to side on the bed, as the patient can be prevailed on to maintain, for such movement of the body increases the peristaltic action of the bowels and the vomiting.

261. While vomiting of choleraic matter continues, there can be no objection against allowing the patient to drink moderately, some say freely, of what he so ardently calls for—cold water; for, whether he drink or not, he will vomit as long as the stomach receives choleraic matter into it, and which ought to and must come away, as very little of it will pass into the intestines and escape by that canal. It is on this account that some practitioners have advised emetics, but of which there is no need, since the morbid matter is an emetic of itself. As the colliquation does not last long, the oozing into the stomach, at first rapid and in great quantity, soon diminishes and but little will remain after the vomiting will have continued for a while; but the irritability it has set up in the stomach will continue when little or none remains. This irritability is so persistent as to force the stomach to reject even the water it so much craves, and which it cannot endure the presence of, even in a very small quantity.

262. It is now time to administer the opiate, to calm a useless irritability which excites efforts to vomit when the need and the good of vomiting has passed away, and which, if allowed to go on, will greatly add to the prostration—the patient already exhausted. Give, then, the opiate in moderate dose, not more than a grain (Sec. 250); if it stay down for only ten minutes it will have, in that short space of time, diminished the irritability, to a proportionate

degree, and tend to its remaining fifteen minutes, perhaps to continue down. Should it be rejected, repeat the dose. When this is well conducted, perfect stillness observed, even of speech, on the part of the patient and the bystanders, one dose, at most, will suffice to calm the organ and soothe the whole system. When this is established—say in half an hour—we may venture to give a tablespoonful of water or tartaric lemonade; in a few minutes more repeat this quantity, gradually increasing the quantity with caution and judgment, so as not to set vomiting once more agoing by the presence of a volume which the stomach will not endure. In a short time more the patient can be safely supplied with as much as is good for him, but never more at one time than three to four ounces at a single draught.

263. A larger quantity of opium than what has been recommended will overdo and paralyze the debilitated stomach beyond a safe point. As soon as it is ascertained that the stomach will bear it, a cup of good broth or beef tea may be added, and in time other appropriate nutriment.

264. *Diarrhœa*, previous to an attack of cholera, or cholerine, as it has been called. It is doubtful whether or not this affection of the bowels is really an antecedent of cholera, or that it is not either an ordinary looseness so commonly met with at all times when no cholera prevails, or in many cases is the result of mental anxiety, disturbing the bowels in a time of great general alarm, may be questioned. This doubt is strengthened by the facility with which the *feculent* discharges can be suppressed and no cholera follow, also by the fact that this diarrhœa may last several days and then cease, even without medication, and that the matter discharged is purely feculent, and has not the least resemblance to choleraic discharges. However, there is nothing remarkable in the fact that some of the cases of " cholerine " should be followed by the supervention of real cholera, since we find the latter attacking patients extremely ill of various fevers during any of their stages, at their beginning, middle or end; such as variola, typhus, etc.

265. In these cases of diarrhœa it will be well to give the patient an opiate, one grain of opium, or thirty drops of laudanum, in a draught of some cordial, spirit or carminative mixture—but not more than one grain or its equivalent. The patient ought to lie down comfortably warm, and remain as quiet as possible. Quietude is particularly called for in every kind of bowel complaint, for exercise or motion increases the peristaltic movement of the intestines, which, of itself, keeps up an irritation and the frequency of the dejections.

These two elements of treatment—the opium and the quietude—will be sure to moderate if not arrest the passages. The practitioner on his next visit to the patient, at the expiration of three or four hours, will find him better, and ought, with a cheerful countenance, to encourage his patient and assure him that his case is under control. The relief that the patient now experiences, and the hopeful assurance of his physician, will calm that anxiety which was a large cause of the bowel complaint.

Should there remain a tendency to purgation still, the same dose may, or rather ought to be repeated, which, with a persistence of repose, moderate warmth, and a few hours of sleep, will produce some twelve or more hours' suspension of the complaint. Should the practitioner, after this interval of relief, judge proper to move the bowels gently, with a view of discharging any of the diarrhœic matter that he may suppose has remained, he may do so by means of a small dose of castor-oil, or a little compound tincture of rhubarb. When this has operated to his satisfaction, he may repeat the anodyne already prescribed, in order to completely calm the intestinal relaxation. There are few cases that will not readily yield to this soothing treatment. No irritating laxative—such as salts, magnesia, senna, or any included under the names purgative, cathartic, drastic—should be permitted to be taken.

266. When the relax has been suppressed the patient may be allowed to leave his bed, should he be desirous of so doing; still, it must be borne in mind that repose of the body gives repose to the intestines. He may be allowed a reasonable amount of sustenance, and rather of that kind he has been habituated to and for which he has an appetite; but thin gruel, weak broths and slops had better be dispensed with. A small glass of any agreeable cordial ought to be taken, or a glass of good wine, such as Madeira, Teneriffe or Port, any *sound* old Spanish wine; but let him shun that villainous compound, now so fashionable, sold under the name of Sherry. This last named wine, even when not adulterated, which is never the case now, is at best a shabby substitute for the warm southern wines already recommended. As regards the light French wines, they are too watery and destitute of *body*, as it is called, to be useful as an auxiliary to the medical treatment of the diarrhœa under consideration. Shun groceries, retail or wholesale, if possible, for neither purity nor truth reside in those haunts of adulteration. Should difficulty be experienced in procuring good wine, dispense with it, and make some kind of palatable cordial or punch, with good spirits.

267. Astounding cases, and cases without discharges, always terminate rapidly in death; consequently they have no sequel. On dissection, no lesion of any kind can be detected; the patient has been simply poisoned.

268. Cases accompanied with rapid and copious discharges, but in which the impulse of the heart is only weakened, the capillary circulation, particularly that portion which belongs to the veins, is still carried on, although feebly, no cyanosis existing, and little or no cramps complained of, will run their course to its termination in a short time, not much over, or even less, than six to twelve hours. Soon after the gastric discharges and that by the skin have ceased, the patient commences to repose, and warmth returns; he will fall into a refreshing sleep, out of which he awakens perfectly restored to health, excepting some slight weakness experienced on exercise, and loss of his usual fullness, both of which are soon restored; he is able to and does go about, feeling well, the next day.

269. Such a case, without premonition, sudden, severe, rapid in its course, termination and recovery, leaving no trace of its fury discoverable in the system, goes far to support the statement, cholera is not a disease; but that the disorder must be attributed to some transient cause—a poison received from without, the potency of which is variable in different persons, astounding in some, severe in others, less in many.

270. There are cases to all appearance of the same degree of severity as described in Sec. 268, and which recover in the same way, but in which debility is more marked, gradually increases day by day until in three, four, or five days the patient sinks without a struggle.

271. Similar cases, which have in addition a slight cyanosis, do not recover so rapidly and completely as 268; but advance slowly in warmth and in convalescence to ultimate perfect recovery. Cyanosis indicates a profound impression made by the poison. These patients require repose, good nursing, a judicious use of cordials, light bitter infusions and careful doses of nourishment. Many of these cases after struggling along for a few days, giving hope, have to yield to a failing economy that has suffered more than it can resist.

272. Severe cases, accompanied with a degree of cyanosis greater than in 271, occasionally weather the choleraic storm, followed by a strong effort of the constitution to convalesce. The pulse is feeble, the temperature slowly returns, the stasis in the skin diminishes in color from leaden to light purple, and in a few days to that of a weak scarlatina. The

eyes are less deeply set in the orbits than they were; the conjunctivæ may appear slightly pinkish; the corneæ are moist, but not sufficiently so to give the clear sheen seen in health, since the lachrymal secretion is scanty, as are all other secretions; urine is formed in small quantity, and strangury has ceased, abdominal contractions no longer pressing the base of the bladder into the meatus. (Sec. 135.) Thirst continues, because repletion of fluid is not yet complete; should the patient sleep with the mouth open, or keep it open while awake, the tongue becomes dry and rough, even brown colored; there is little appetite excepting for fluids, a little of which, with the nutriment that happens to be mixed with it, may pass the pylorus in some degree digested, perhaps, and which will furnish material for a scanty stool, not much affected with bile. The temperature of the surface is low, but as the skin is dry and cannot evaporate, it gives the sensation of a degree of heat that does not exist in reality; neither is this sensation of heat, nor the thirst, nor the parched mouth, due to fever—there is none. I closely watched several cases, particularly those of two ladies. The mind was slightly blunted, it is true, and like in fevers the patients looked slightly anxious, but were quite clear in mind to the last. They were rather restless, soon weary of one position; one would get up of herself,

10

sit on her chair for a while, say but little, and return to bed; none complained of that false heat so constantly present in sensation during colliquation. One of these died on the 10th day, with her senses, as did the other on the 12th. Some such patients gradually recovered, much deteriorated in constitution for some months.

The treatment of alleged typhoid state ought to be principally expectant. Husband the strength of the patient by avoiding any thing that can fatigue, waste, or exhaust the patient. Give such nutriment as the appetite prefers, and in such quantities as the stomach can master, not more. Palatable fluids, and now, light French wines freely, unless disagreeable to the patient; in this case give cordials at judicious intervals. Tepid baths, when the patient can get into them without fatigue; in their stead, sponging frequently. This external wetting will contribute to supply the watery waste that has taken place, will aerate the cyanosed surface, and prove a grateful anodyne to the feelings. Never give a pharmaceutical stimulant when a natural one, like wine, can be had. Shun medication, so-called.

www.ingramcontent.com/pod-product-compliance
Lightning Source LLC
Chambersburg PA
CBHW021707210326
41599CB00013B/1553